CONTENTS

·PREFACE·

My Shelter-Building Adventures

My greatest feeling of accomplishment as a kid occurred the day I sat in an old wicker chair outside my newly completed clubhouse. We—two girls and two boys—had just finished nailing up shingles we had scrounged from a building site and hauled home in a wagon. I was beaming with pride at what we'd achieved, having built this entire structure ourselves without any help from grown-ups.

This was the third clubhouse that I had built. What drove me and my friends to do this, I believe, was a basic desire to have a place, a room, a shelter of our very own that was away from the scrutiny of parents. My bedroom at home was comfortable, of course, but it just wasn't *mine*.

It all started when I'd become enchanted by a book in the school library titled *Billy's Clubhouse*, and also by the stories that my mom used to tell me about her brother, Frank. Apparently, Frank had lived in a shack he'd built for himself across the street from their home (I assume with their parents' blessing!). If he—and fictional Billy from my library book—could do that, then why not me?

Building a shelter became a mystery that had to be solved. In our Southern California neighborhood, I wandered through houses under construction trying to decipher the logic of how they were put together. This eventually led me to Mr. Barnes, a kind neighbor who was building his own house by himself. Instead of chasing me off (as some might do), he seemed genuinely amused by my curiosity. Every time I'd stop to watch, he'd pause what he was doing to show me how to use his tape measure, handsaw, and other tools. To an 11-year-old kid without a dad, these weren't only revelations about construction, they were also signs of acceptance and encouragement from someone I looked up to. I soaked it up!

As fate would have it, I was given a rare opportunity to build my own bedroom when I was 16 years old. My mom and I had moved to the far reaches of Northern California where she was to be the property caretaker for her in-laws, Jesse and Louise. They provided us with a cottage to live in, but there was one big hitch—there was no bedroom for me! Jesse had generously offered to supervise me if I wanted to build a backyard bedroom, but *only* if I agreed to do the actual carpentry myself. I was both honored and humbled, since he had built their entire house and garage as well as the cottage. Jesse was often abrupt and had little patience if I bumbled about or asked too many questions, but what I learned from him was invaluable. By the time school had started, my room was finished, complete with a wood-burning heater.

During the next few formative years, I pursued my other passions of art and architecture at college. But my first love—construction—remained a constant fascination, and I would often build models of homes or draw house plans in my spare time.

This was all happening during the difficult days of the Vietnam War. Not only was I going broke and failing my classes after navigating different paths with my studies, but the local draft board also had me in their sights, so I decided to get my military obligation over with. I left school, enlisted in the US Army, and was shipped off to fight in the war—but in a stroke of luck that I'm forever grateful for, I was reassigned as a company clerk at the last minute. During my free time, I happened upon a house-building project in the village I was stationed in. A local Vietnamese family was constructing their entire house frame with just mortise-and-tenon joints (no nails) and using only hand tools. Once again, I was enchanted!

After returning to the States, I picked up where I left off with my art studies but soon drifted back into carpentry to pay the bills. With a beat-up VW bus for hauling tools and materials around, I first painted homes and repaired porches and decks, then moved on to room additions and eventually cabins—including my own—all in Northern California.

I was quickly realizing that all I really wanted to do was design and build small, self-contained homes, similar to the compact, shingled cottages that I'd often admired as a kid in my old beach-town neighborhood. After I graduated from art school, my construction dreams reemerged when I discovered the relatively unsettled Oregon Coast. I had found a vacant lot just two blocks from the ocean in Pacific City, and, while saving money for materials, I spent two years dreaming of what I might eventually build upon it.

This was to become my *ultimate* clubhouse—an actual two-story home with a kitchen, bathroom, living room, and single bedroom. I built it with used or odd-lot materials due to my itinerant-carpenter budget. (A bank loan was out of the question!) Only the framing lumber, plumbing pipes, and wiring were new, as per building-code requirements. The house took another two years to complete, and during that time, I found a like-minded community of other do-it-yourself home-builders scattered around the area. (Or, I might say, they found me! Some of those friendships are still going strong now, 40 years later.)

In a twist of fate, I also met my wife, Vicki, when my house was nearly finished. We bonded (at least partly!) because of her willingness to help me enlarge the living room, from digging the foundation trenches to pounding *a lot* of nails. We're still going strong today, but now living in Wisconsin where we have extensively remodeled and added to our Victorian-era house while raising our three daughters. Our oldest, Kate, is now helping me design a kind of "granny cottage" for her in the backyard, which we'll break ground on next spring.

Building a backyard shelter is a joy that never ends, and one that I'm eager for you to experience as well!

There's nothing like a place of your own . . .

·INTRODUCTION·

You Can Build It!

Do you need more living space? Are your kids asking for a playhouse? Do you have a dependent relative who needs a room of their own, perhaps just a few steps from your back door? Or do you simply need a quiet space in which to write, compose, paint, create, or just sit and be? If you have some extra yard space and local zoning allows for an additional structure, there are so many possibilities!

There are countless YouTube videos and guidebooks that demonstrate every facet of house-building. It just takes a little time and patience to learn how to measure and safely cut a board, pound nails without bending them, and figure out which pieces go where. This book is intended to get you beyond the how-to videos and into the lumberyard with a materials list in hand.

Few things are more rewarding than building your very own custom shelter. Call it what you will—a granny cottage, she shed, man cave, clubhouse, or garden shed—it can be your own special place that *you* created.

Building can sometimes involve tedious work, but the reward is immeasurable. Getting some help to lift and carry heavy materials is often essential, so you might hire an occasional helper or, better yet, ask a family member who will benefit from the project to pitch in. I've found that most kids love the idea of helping to build a space of their own, although they might not always be excited about the actual grunt work! But participating in even a small room build can have a profound effect on its potential occupant.

Sometimes the most fun (and daunting) part is designing your shelter. How big should it be? How tall? Will it fit everything you want? You can have fun with this phase because all it takes is your time, but it is time well spent. Write down all the possible uses you can think of for your space. For instance, if it's a music studio, what instruments will you want in it? Will larger instruments, like a keyboard, be added later? Where will you store all the mics, cables, and sheet music? Where will the outlets be to power everything up? Do you want a snack counter with a microwave and a mini fridge? Will you want lots of daylight, or some artificial lighting you can control for music videos? Such a list can be essential to help you envision the shape and size of the space that you will need.

The following pages include some ideas and designs for a variety of backyard structures, from a sturdy 6 × 8-foot hut to an 8 × 15-foot palace. You can always expand upon these designs or add more rooms to fit your needs later. However, keep in mind that any structure—and especially larger structures—may be subject to the local building code and zoning laws, so first check with the local building department to learn about rules and requirements that apply to your project. You may need to obtain a building permit, although this (usually) is not that difficult.

From Classic to Complex

Here are some design ideas to get your creative juices flowing. From a simple one-room shelter to a sprawling, multilevel space, you can build whatever you want, though you might want to start small while you hone your skills.

The Classic Design

The Classic Design is a simple, good-size space for most purposes. I'll explain how to design and build this structure in Part 2.

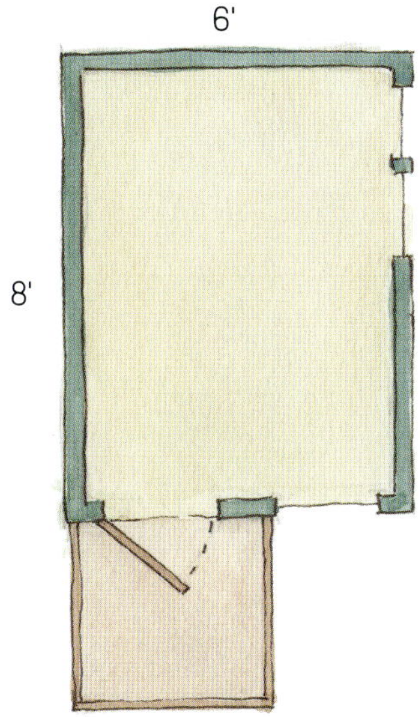

6'

8'

Many variations are possible. The doors and windows can go in different places, and you can decorate it however you want, inside and out.

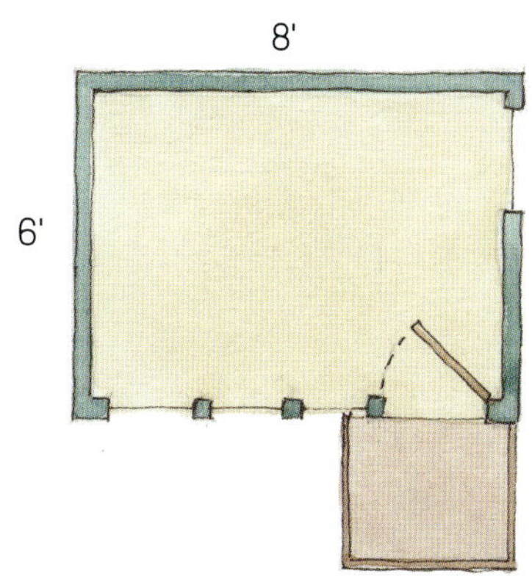

Add a Patio

This is how a Classic Design structure might look with a brick patio and a trellis built out of 2×2s supported by two 4×4 posts and a 2×6 beam. Canvas, split bamboo fencing, or old bamboo blinds draped over the 2×2 framework would add more shade, if desired.

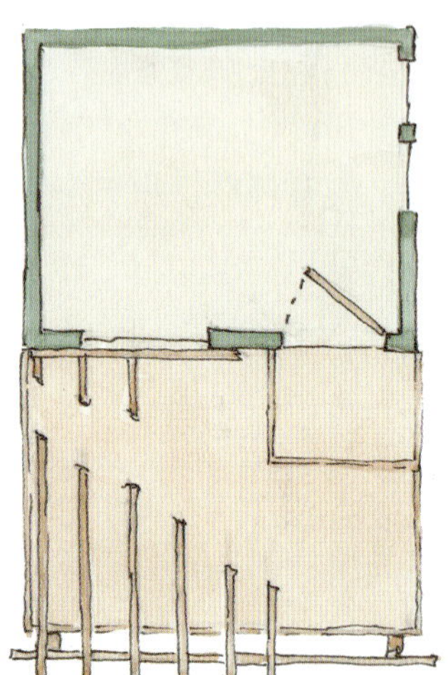

The El

This is a Classic Design with a room added and with the door and windows in different places. (See Chapter 6 for details on building a room addition.)

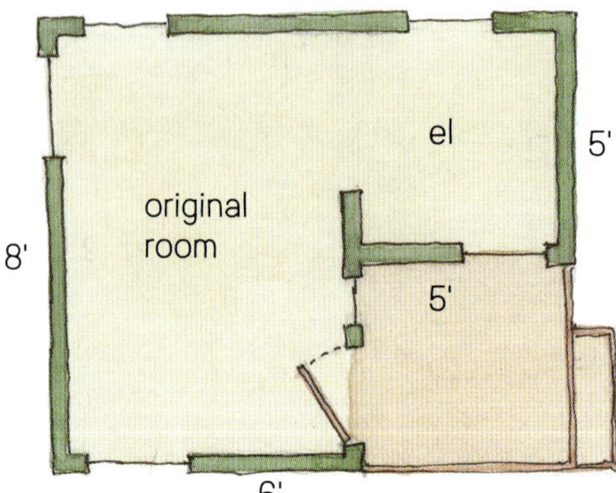

BEGINNER'S GUIDE TO Backyard Building

SIMPLE STEP-BY-STEP INSTRUCTIONS FOR CONSTRUCTING YOUR OWN Shed • Studio • Tiny House

Storey Publishing

The mission of Storey Publishing is to serve our customers by publishing practical information that encourages personal independence in harmony with the environment.

EDITED BY Carleen Madigan, Lisa H. Hiley, and Christine Maxfield
ART DIRECTION AND BOOK DESIGN BY HK Goldstein
COVER AND INTERIOR ILLUSTRATIONS BY © Lee Mothes, except page 16 which is excerpted from *Audels Carpenters and Builders Guide #4* by Frank D. Graham and Thomas J. Emery (Theo. Audel & Co., 1923)
PHOTOGRAPHY BY © Lee Mothes

A version of this book was previously published under the title *Keep Out!* by Lee Mothes (Storey Publishing, 2013).

Be sure to read all of the instructions thoroughly before undertaking any of the projects in this book and follow all of the safety guidelines. Take proper safety precautions before using potentially dangerous tools and equipment.

Storey books may be purchased in bulk for business, educational, or promotional use. Special editions or book excerpts can also be created to specification. For details, please contact your local bookseller or the Hachette Book Group Special Markets Department at special.markets@hbgusa.com.

Storey Publishing
210 MASS MoCA Way
North Adams, MA 01247
storey.com

Storey Publishing is an imprint of Workman Publishing, a division of Hachette Book Group, Inc., 1290 Avenue of the Americas, New York, NY 10104. The Storey Publishing name and logo are registered trademarks of Hachette Book Group, Inc.

ISBNs: 978-1-63586-960-6 (paperback);
978-1-63586-961-3 (ebook)

Printed in China by Toppan Leefung Printing Ltd.
on paper from responsible sources
10 9 8 7 6 5 4 3 2 1

TLF

Library of Congress Cataloging-in-Publication Data on file

To Virginia, Kathy, Chris, Chele, Jesse Manley, and Mr. Barnes. And to Mom, for enabling me to build that first clubhouse and everything after that.

A Tiny House with Many Rooms

If you have the space and your loved ones all want to have a nook or alcove of their own, this is possible. Starting with a Classic Design, you can add rooms to the outside as you need to. In one of them, you can build a secret escape hatch for adventurous kids. You might also have space inside to add a partition. Just remember to build the added roofs with a slant so they don't leak (that would put a damper on any fun get-together!).

The Unconventional Building

People have built structures with bottles, cans, tires, newspaper (mixed with cement), cordwood, straw bales, and canvas. Others have used old boats, shipping pallets, and shipping containers. There is at least one house out there covered with license plates and another with a porch shingled with phonograph records. I once saw an entire floor built with old plywood road signs discarded by county highway departments. With a little imagination, you can use almost any materials for your backyard building instead of buying all new stuff (plus it's better for the planet and your savings account!).

The Fun-with-Doors Design

If you can find about 6 to 10 solid-wood doors (not hollow-core or particleboard-core), you can use them to build a unique structure. Salvage yards sometimes sell old doors quite cheap. Doors with glass in them can also be used as windows. With a clubhouse made of doors, your inner circle can be the only ones who know which door actually opens (a fun party trick!).

The Pile

A backyard building can evolve into something much different from your original design. The materials you find and location you wind up in can challenge your flexibility, and that can be a good thing: The result can be truly original.

It's best to start with a plan, but it's okay to modify that plan if you find six old doors or you simply get a new idea. You are the master of your design! It can be challenging to change your design to fit something new you found, so just relax and take your time. This is *your* space for applying your own creative vision, and it's all right to make mistakes.

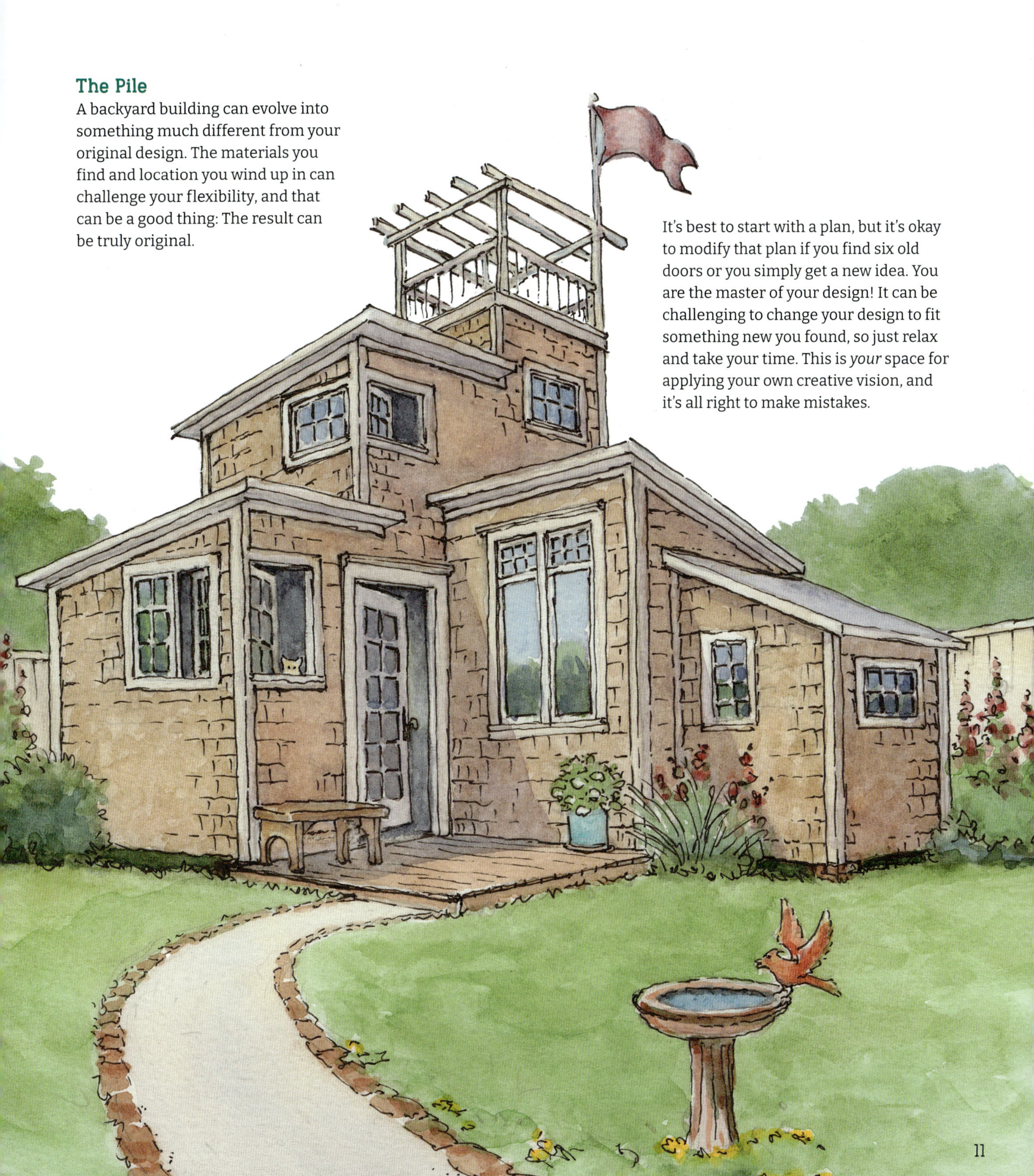

GETTING READY TO BUILD

Backyard designs range widely. Much will depend on the materials you find and on your own ideas. Simple is good, especially in the beginning. The best way to start out is by building a one-room structure, as described in Part 2. To this classic clubhouse design you can add a room, a garden patio, or maybe a second story. You can also add interior walls (partitions) and do all kinds of things to the inside and outside to make it as comfortable or as unconventional as you'd like.

This section introduces you to the essential tools you'll need to build your structure, discusses where and how to find the materials for it, and helps you figure out a design.

·CHAPTER 1·

Tools and Techniques

A backyard dwelling is essentially made of wood and nails, and to build one, all you need are a few common hand tools. In this chapter, I show you how to use these tools. For more experienced builders, I've added a few make-life-easier options, including some basic power tools.

If you've never done any building before, be patient with yourself as you learn how to hit nails and saw boards. You're learning with your body as well as with your mind, and, as with dancing, tennis, or snowboarding, that takes a while. Once you learn the basics, you'll find it relatively easy to master other carpentry tools and to build bigger or more complex structures.

tape measure

pencils

screwdriver

level

crowbar

claw hammer

handsaw

The Nine Essential Tools

The first thing you'll need before building anything is a set of simple, good-quality hand tools. Believe it or not, you can build an entire house with just the nine essential tools listed here. Having the right tools saves you time and money in the long run.

1. **A 16-ounce claw hammer.** This is a finish hammer, which is lighter than a framing hammer and just right for new builders. A wood-handled hammer is fine, but a fiberglass-handled one will last longer.
2. **A 15-inch-long handsaw.** This is often called a toolbox saw.
3. **A tape measure.** A 12- or 16-foot-long tape is fine for smaller structures, while a 20- to 25-foot-long tape is more suitable for ambitious designs.
4. **A 7-inch rafter square or speed square.** Use this for marking boards to make accurate saw cuts. Although shaped like a triangle, it's called a *square* because it has two sides that make a right (90°) angle, which defines a square in geometry. Later, I'll show you how to use the marks on this kind of rafter square to cut boards at angles.
5. **An ordinary pencil or a carpenter's pencil.** You'll use this for marking boards.
6. **Screwdrivers.** You can get a Phillips-head screwdriver (which has a cross-shaped tip) and a standard straight-slot, or flathead, screwdriver, or you can get a combination-style screwdriver with interchangeable blades. The combination screwdrivers include a set of smaller blades, but they have one drawback: The small snap-in blades can easily get lost. Make sure the blades or tips are made of hardened steel; cheap ones wear out fast.
7. **A level.** Look for a 9-inch-long plastic "torpedo" level. These are quite accurate and can fit in any toolbox.
8. **A crowbar.** You'll need this to remove nails and pull apart mistakes (they happen!).
9. **Sawhorses.** Sawhorses make building much safer and easier. The plastic fold-up kind are sturdy and a good height for experienced builders with power saws. Lower sawhorses are better for hand sawing and for new builders. You can assemble your own sawhorses with some common boards and these essential tools (see how on page 20).

One more sort-of-tool: A nail pouch or nail apron to hold a tape measure, pencil, and nails will be useful. Lumberyards sometimes give these away.

sawhorses

Safety Means Taking It Slowly

Before you begin, take a deep breath and relax. Building *always* takes longer than you think, especially if this is your first project. Nails get bent, saws cut crooked, thumbs get whacked, and splinters get under your skin. Most of these annoyances and injuries are the result of not focusing or trying to do something too fast. Regroup and then start again. Here are a few things to remember about safety.

- Focus only on sawing when you saw a board. Find a helper to hold the board down, or use clamps, and saw slowly. Stop sawing if you want to talk.
- The same thing applies when you pound in nails: Focus only on the nail and where your hands are.
- Don't leave nails sticking out of boards! Some barefoot person is bound to come around and step on the one board in the whole yard that has a nail sticking out. Pound them back through, pull them out, or bend them over flat. This also applies to nails that are poking through the inside of your structure's walls.
- Wear gloves if you are handling recycled or rough-sawn lumber. Wood dries out over time and can get splintery.
- Wear safety goggles and a dust mask when you're using a power saw. Also, unplug power tools if you are going away, even for only a few minutes.

Pounding a Nail

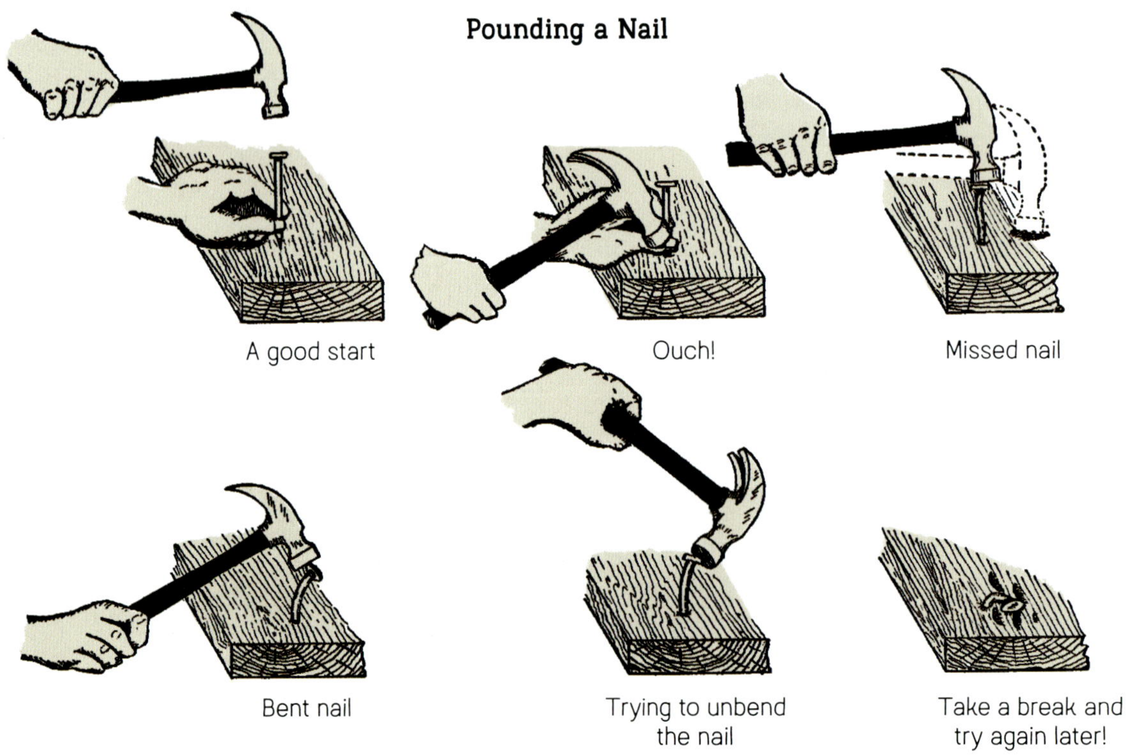

How to Pound a Nail and Measure and Saw a Board

Before you get started on your first building project, you'll want to practice pounding nails, using a tape measure, and cutting a board. Think of this as rehearsal.

POUNDING NAILS

You've probably pounded a few nails in the past, but let's pretend you're doing it for the first time. Get your hammer, a few large nails, and a long, thick board like a 2×4, and get comfortable.

If you are right-handed, hold the nail with your left hand and the hammer with your right. If you are left-handed, do the opposite. Hold the hammer handle at the far end, away from the hammerhead, and tap the nail into the wood hard enough so it stands up on its own. Then move your nail-holding hand out of the way and drive in the nail. The drawings on page 16 show the right way to start and what can happen if you miss. Just keep trying.

MEASURING A BOARD FOR CUTTING

To mark a cutting line on a board, you'll need a tape measure, pencil, and square. Hold these tools in your nail apron or tool belt.

All tape measures have numbers and small marks that show feet, inches, and fractions of inches. (Some tape measures also show centimeters and meters, but we'll only use inches here.) The numbers tell you the feet and inches, and the small marks show the fractions of an inch, including halves, quarters, eighths, and sixteenths.

Get a board from your lumber pile—say, a 2×4. To find out its exact dimensions, hook your tape measure across the wide part of the 2×4. You'll see that the board is actually about 3½ inches wide.

Unbending a Nail

Hook the claws just above the bend.

Then twist the hammer sideways. This usually straightens the nail, but not always.

Prying Up a Bent Nail

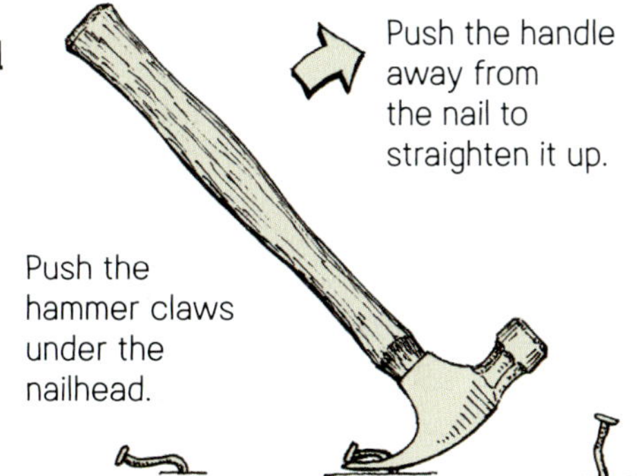

Pulling a Nail

If the nail is hopelessly bent, it's not worth the effort. Pull it out, using a block of wood for leverage, as shown.

Reading a Tape Measure

It shows feet, inches, and fractions of inches.

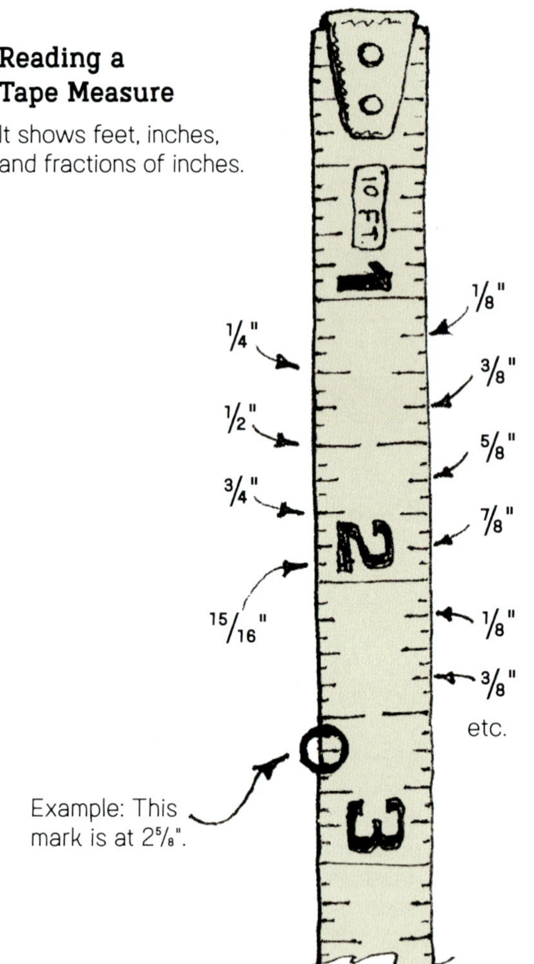

Example: This mark is at 2 5/8".

(Why is it not 2 inches by 4 inches? See the box below.) Next, measure the board's thickness; you should get about 1½ inches. As you build with these boards, you'll be using these actual dimensions in your planning.

Now let's mark the board to prepare it for cutting. Say you need some 6-foot-long 2×4s to support your floorboards (which you will), and you have 8-foot-long 2×4s in your pile. Put one of these 2×4s on your sawhorses (or chairs or a bench). Hook the tape measure at one end and pull the tape down the board until you reach 72 inches, which is 6 feet. The tape measure might say "6F" there, too, which also means 6 feet.

Using your pencil, make a mark on your board next to that mark on the tape measure, right between the 7 and the 2 in 72. Then set your square on the board, with its wide, or flanged, edge pulled tight against the bottom of the board. Set the square's other edge to your mark and draw a *cut line* over your mark with your pencil, as shown on the next page.

"2×4" Is a Name, Not the Actual Dimensions

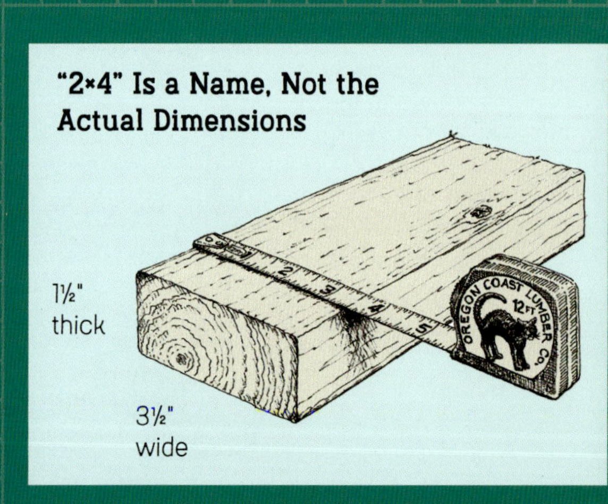

Why a 2×4 Isn't Really 2" by 4"

At the lumber mill, logs are rough-sawn into boards that are 1 or 2 inches thick and 2 to 12 inches wide. The rough boards are then surfaced and smoothed with a planer, which makes the boards smaller. Therefore, a so-called 2×4 might have started out at 2 inches thick and 4 inches wide (its *nominal* size) but has been planed down to 1½ inches thick and 3½ inches wide (its *actual* size).

Older salvaged or unplaned boards might be closer to their nominal dimensions. Try measuring different boards to see what you get.

SAWING A BOARD

Now that your board is marked, get your saw and sawhorses if you have them. (To build your own sawhorses, see page 20.) To start, see the drawing below to figure out how to hold down the board. You can have a helper hold the board for you, or use clamps if you have them.

Use a sharp saw. It will make life so much easier. Start by setting the teeth of the saw at the cut line, then pull up the saw to start the cut (see the drawing below). Saw with long, slow strokes. Leaning over your board as you work (careful to keep loose clothing out of the way), push and pull the saw along the line you drew. You'll find that the push strokes do most of the cutting. If the saw wanders off the line, back up and slowly, patiently resaw toward the line. After a while, you'll be able to saw on the line you're following with ease. When you get near the end, saw the last stroke fast so the scrap falls off without taking a splintery piece of your good wood with it. In the meantime, it's okay if the cut is off the line a little; don't worry, this is still batting practice.

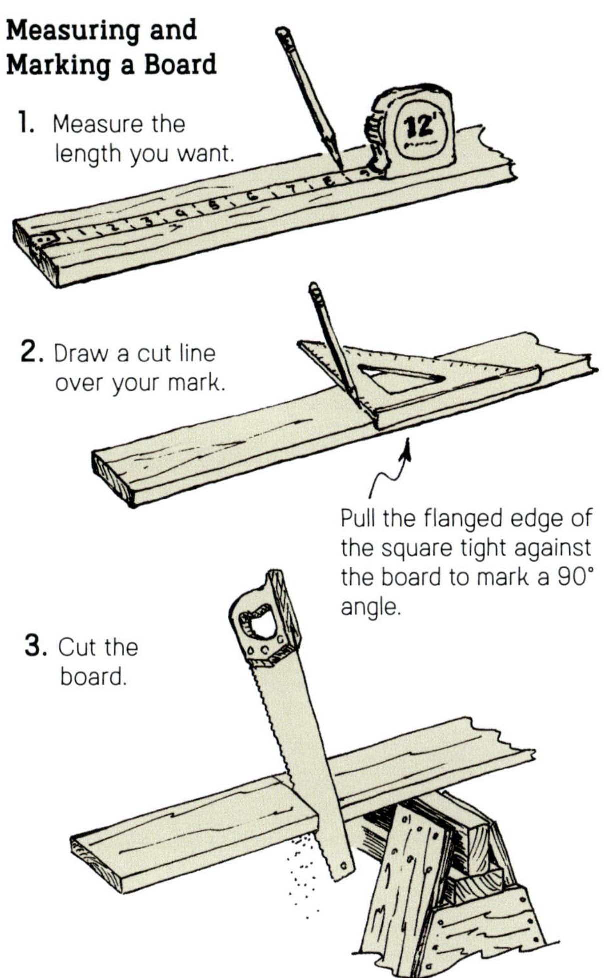

How to Build Your Own Sawhorses

You will need some 2×4 and 1×6 boards, some scraps of ½-inch-thick plywood, and nails. Have your tape measure, pencil, saw, square, and hammer handy. Having a bench or other sturdy surface to cut wood on is helpful. Here is a list of what you need to make two sawhorses.

SAWHORSE MATERIALS

QUANTITY	DESCRIPTION
2	2×4s, 6 feet long
2	1×6s, 8 feet long
1	2 × 4-foot panel of ½-inch-thick plywood (or one 1×6, 4 feet long)
1 pound	6d coated sinkers or galvanized box nails
1 pound	8d coated sinkers or galvanized box nails
1 pound	12d or 16d coated sinkers or galvanized box nails

Homemade Sawhorses

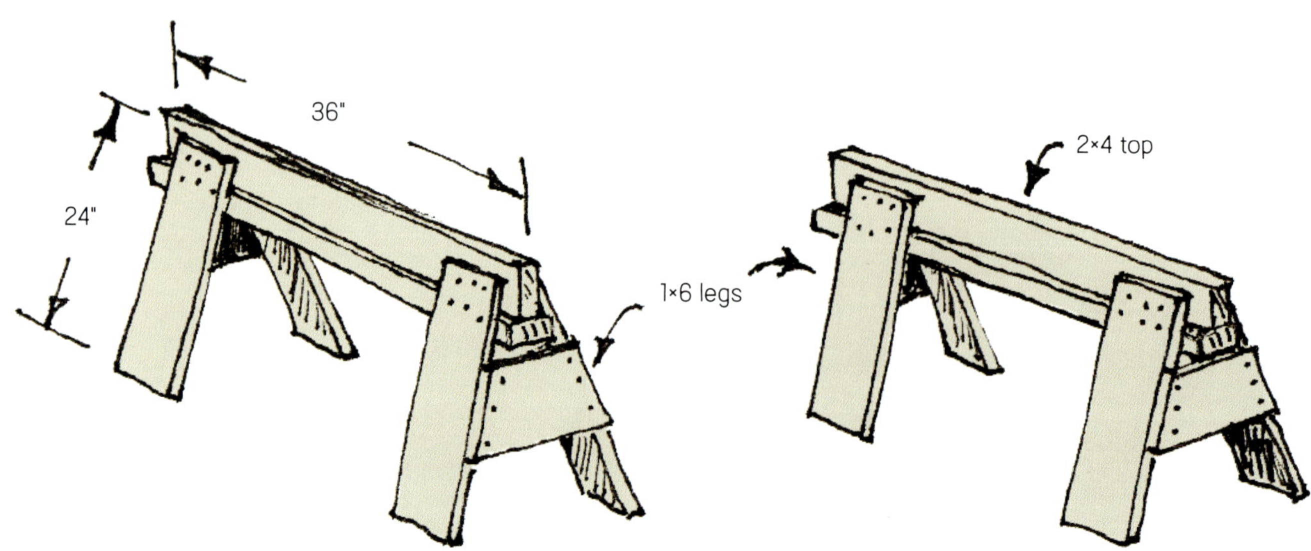

Step 1. For the sawhorse tops, cut four pieces of 2×4 that are each 36 inches long. To do this, measure to 36 inches from the end of each 2×4 with your tape measure. Mark the spot with your pencil, then use your square to draw a cut line at the mark. Saw along the cut line with a sharp handsaw.

Step 2. Find a hard floor or concrete surface on which to nail these together. Nail the 2×4s together with 12d or 16d nails, as shown in the drawing at right. Use four or five nails for each pair.

Step 3. For the legs, cut eight 24-inch-long pieces from your 1×6 boards. Nail these to the 2×4 tops, using five or six 8d nails per leg.

Step 4. For the braces, hold a 1×6 board or a small piece of plywood against the end of the sawhorse, as shown at right. Draw lines where this board (the brace) covers the legs, then saw off the extra wood. Using three or four 6d nails per leg, nail the braces to the legs.

Nail the Tops Together

Nail the Legs to the Tops

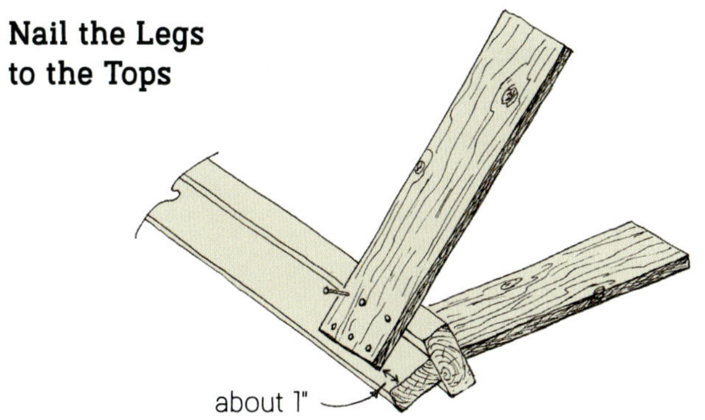

Put on the Braces

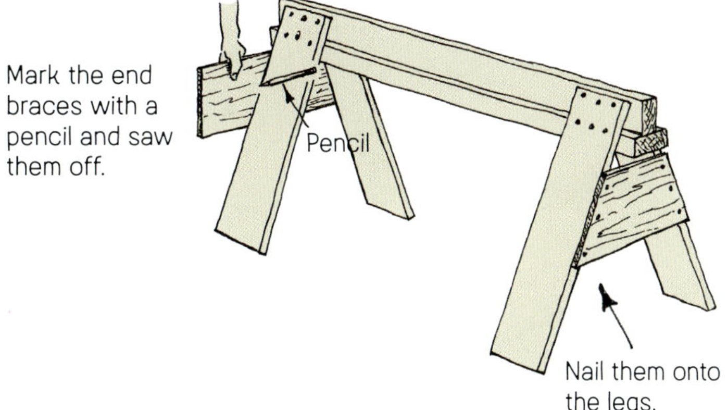

Building sawhorses is good practice and also kind of a test. If you hated every part of doing this, then now is the time to decide whether you really want to build your own backyard dwelling!

Make-Life-Easier Tools

As you gain practice using your nine essential tools, you'll eventually feel the need for more tools to make building easier and safer. These tools will also be useful to put siding, roofing, and trim boards on your structure.

- **Clamps.** If you feel your board is wiggling too much when you're sawing, and all of your helpers have left, you might want clamps to hold it down. Clamps are also great for gluing and other assembly. The modern, lightweight quick-grip clamps are much easier to use than the traditional iron C-clamps. I recommend a set of two 12-inch clamps or two 6-inch clamps.
- **Pliers and wire cutters.** A basic set includes standard pliers, wire-cutting pliers (great for biting off small nails that won't pry out), and needle-nose pliers for all sorts of uses.
- **A utility knife with a retractable blade.** Use this knife for cutting tar paper or shingles for your roof, making cut lines for cleaner saw cuts, or shaving wood that's too thin for a saw to cut off. Be careful—the blade is sharp!
- **An awl or scratch awl.** This is used for poking pilot holes for screws or for scratching a cut line if you don't have a pencil. An awl is also handy to hold the end of a tape measure to a piece of wood.
- **A nail set.** To sink nailheads into wood, use a nail set, which you hit with a hammer. A medium size with a $\frac{3}{32}$-inch-wide point is good for most needs.
- **A ¾-inch-wide wood chisel.** The steel cap on the chisel allows you to hit it with a hammer, which is necessary for most chiseling. Watch for nails when chiseling—the hardened-steel blade will chip easily.
- **A small hand plane or block plane.** Planes are helpful for shaving wood, especially when fitting windows and doors. Make sure you don't hit any nailheads while using the plane—it's very bad for the blade!

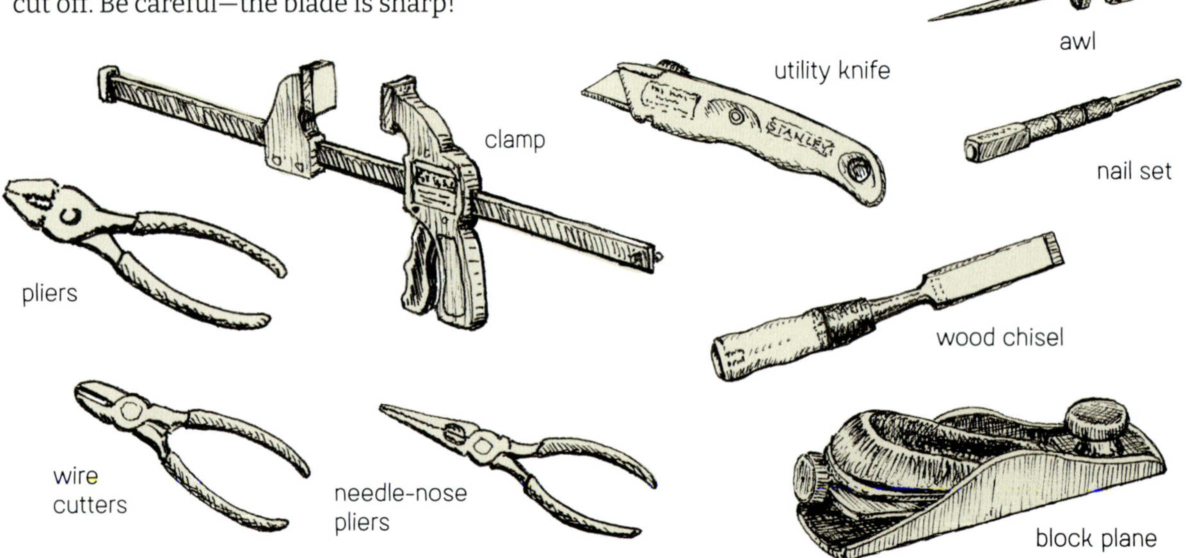

- **A sanding block and sandpaper sheets.** Use these for knocking off splinters and smoothing wood before painting. Get 100-grit, or medium-grit, sandpaper to start with. Grit size is roughly the number of sand grains per inch. The lower the number, the bigger the grains. Sandpaper at 400-grit is very smooth, while 50-grit is rough.
- **A toolbox.** You'll want one that is at least 19 inches long to put all your gear in.
- **A longer level or a straight board.** Once you start building bigger floors and higher walls, you'll want a longer level. An alternate trick is to attach your torpedo level to a straight 6- or 8-foot-long 2×4 with rubber bands. Cut notches into the board where the rubber bands are placed so the board rests flat while you are leveling (see drawing on page 45).
- **A 6-foot stepladder.** Someone probably has one you can borrow if you don't want to make the investment just yet.
- **Plumb bob and layout line.** A plumb bob is a great tool for finding out whether something is "plumb," or straight up and down. Layout line, sometimes called mason's line, is strong white or colored string that usually comes on a handy plastic spool. These tools are both used to lay out a foundation on bare ground. An 8-ounce brass plumb bob is best. To make it work, attach a 9-foot-long piece of the layout line inside the removable top of the plumb bob.
- **Chalk line.** A chalk line is a long string coated with blue or red chalk that is used to make long saw-cut lines or guide lines, such as when nailing shingles on a wall. The line is kept in a chalk box that you fill with powdered chalk from time to time.

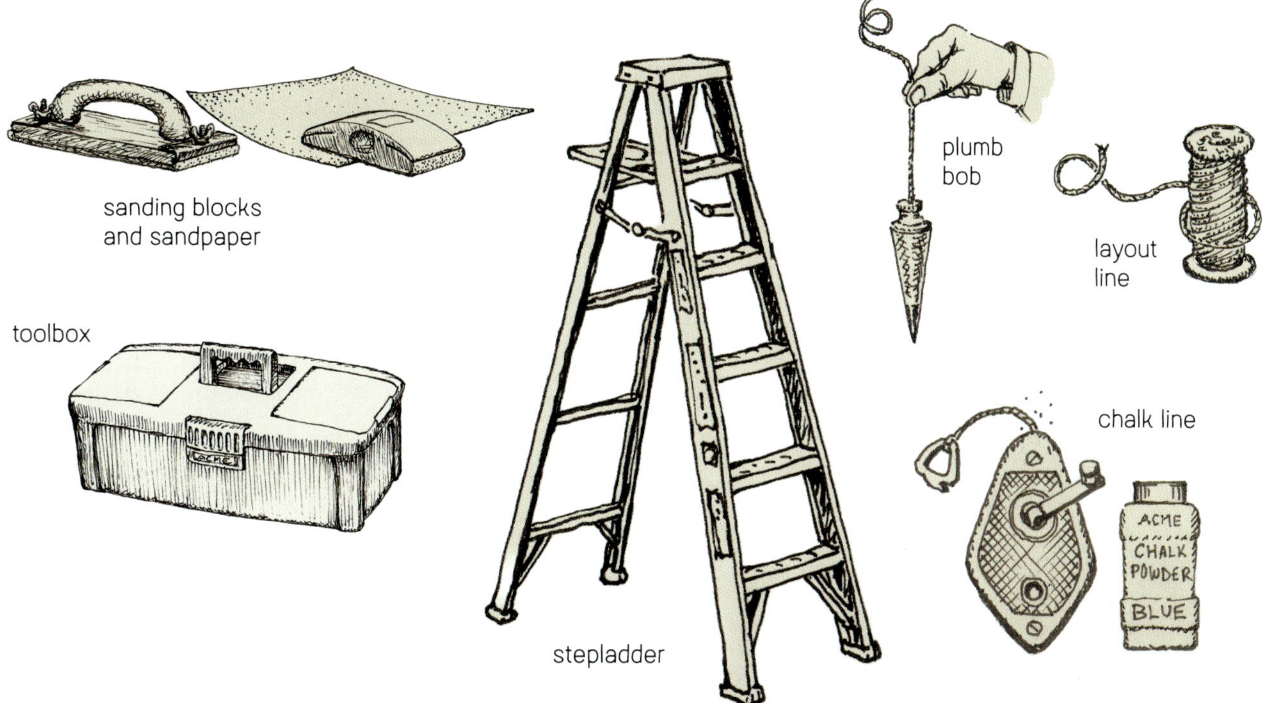

- **Line level.** This 3-inch-long level is designed to hang on a tightly stretched layout line. It will give you an accurate enough reading to level a small foundation.
- **Framing hammer.** A framing hammer is relatively heavy (at 20 or 22 ounces), which makes it easier to pound in all those 12d or 16d nails (I'll explain nail sizes in Chapter 2). Get one with a smooth face rather than a waffle face; a waffle head chews up the wood (or your hand!) if you miss.
- **Sledgehammer.** Get one with a 3- to 6-pound head to drive in stakes and for other blunt pounding chores. Choose the heaviest one you can comfortably swing.
- **Long, wide tape measure.** If you're building a large structure, you'll want a longer tape measure, in the range of 20 to 25 feet long. Get one with a ¾-inch- or 1-inch-wide blade, which can be extended farther without bending.
- **Caulk gun.** This "gun" holds a variety of construction adhesives used for gluing sheathing panels to the shelter frame, as well as caulk for sealing narrow gaps between siding and windows. To use it, put a tube of caulk in the gun, trim off or open the tip of the caulk, then squeeze the trigger of the gun until the caulk comes out. To stop the caulk, release the tension on the gun's plunger (sometimes you have to pull the plunger back, too).
- **Staple gun.** This will come in handy when you are putting up plastic vapor barriers, roofing underlayment (tar paper), and insulation. Try out different models to find one that feels comfortable. Some heavy-duty staple guns require a powerful grip and can get tiresome to operate.
- **Tool belt.** A carpenter's tool belt certainly makes life easier. The best "system" I know of consists of a nylon web belt, a couple of bags, and a steel hammer loop. One bag is for nails, and the other is for small tools such as your square, tape measure, pencil, chisel, and so on. Buy the bags, loop, and belt separately, and then adjust their position to your liking.

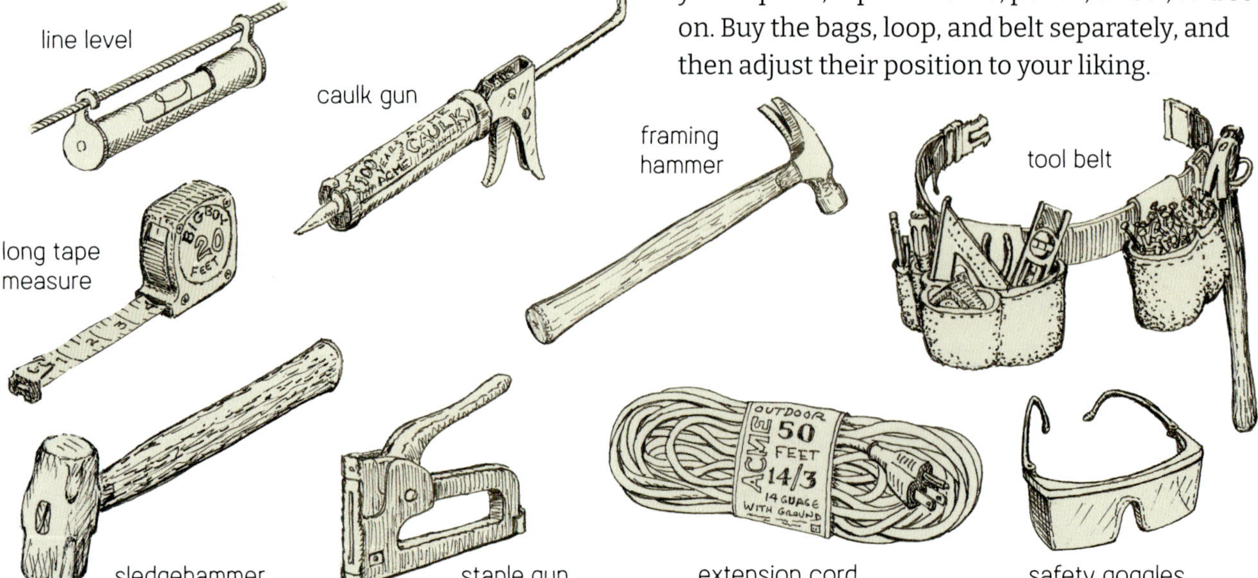

Power Tools

Once you have practiced with the hand tools you've acquired, you'll likely want to get a few basic power tools to save time. A portable circular saw, a power drill, and a jigsaw or saber saw are all you really need.

A portable power circular saw can be intimidating at first. It's loud, it jerks when the motor starts up, and it has a sharp and fast-moving blade. But fear not—the saw is safe when handled properly. If you've never used one before, read the user's manual for your saw, especially the parts about how to hold the saw and how to adjust the blade for different board thicknesses. Also have someone experienced show you how to use it before you take over yourself.

When using a power saw, always cut a board so the scrap falls off outside of your sawhorses, never between them, as shown below. If you cut the board between the sawhorses, it will pinch the blade and the saw might kick back at you. While cutting, push the saw gently and slowly across the board—let it do the work for you. The saw throws sawdust all over the place, so wear safety goggles.

Power drills are not only for drilling holes but also for assembling entire projects with construction screws instead of nails. A variable-speed reversible (VSR) drill works well as a power screwdriver. A battery-powered one has the advantage of being cordless. A standard VSR drill is designed for precision when drilling holes and driving screws, but for bigger jobs, a more powerful impact driver-drill may be the better choice.

A power jigsaw (also called a saber saw) is great for cutting out holes or making curved cuts on plywood panels or any wood up to 1½ inches thick.

A heavy-duty electrical extension cord with 14-gauge or 12-gauge wire with a ground is required to run power tools. A wimpy cord might damage the motors in your power tools. The ground is the third prong on the plug; it protects you from electrical shocks if the tool gets wet or the cord is accidentally cut. Get a cord that is 25 to 100 feet long, depending on how far your job is from the nearest outlet.

circular saw

power drill

drill bits and drivers

jigsaw

Using a Circular Saw

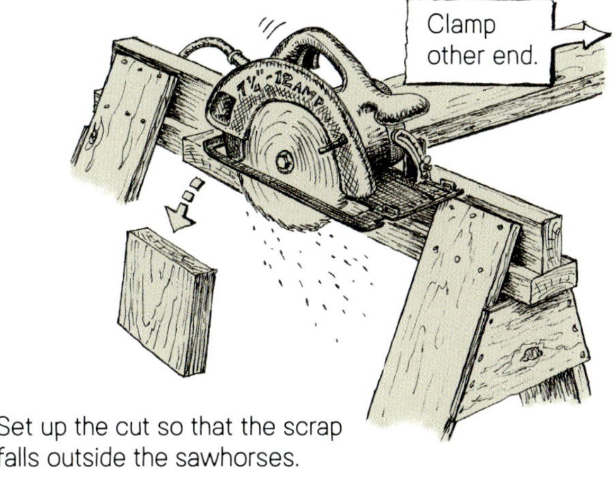

Set up the cut so that the scrap falls outside the sawhorses.

· CHAPTER 2 ·

Making Plans and Finding Materials

You've assembled enough tools to get started, and now you are ready to make your construction dreams a reality. Next up is finding a building site and deciding how big you want your structure to be. Then you'll acquire the materials.

Finding the Best Building Site

A good building site offers adequate room for access around all four sides of your structure and perhaps room on one side for a patio, deck, or garden, if you decide to add on later. Consider how to orient your structure, too: Check out where the sun rises and sets—perhaps you'd like morning light for breakfasts or you're planning to watch the sun set as you wind down from a busy day.

Be careful not to choose a site too close to a neighbor's house. If there is no fence, talk to your neighbors about where the boundary is, or take a look at your home's deed. (If there is a fence, don't use it as one of your walls!) Furthermore, don't build in front of someone's picture window or otherwise block their view—you will get complaints. Also, if your house is nearby, will your new construction look appealing, or will it be in the way of a view?

When you have a good spot in mind and have determined the size of your structure, make sure your building will fit where you want it to go, especially when considering any setback rules (see the next section). Mark the actual dimensions on the ground, measuring with a tape measure and then laying down boards or a rope to outline the space. Imagine where the door will be and how it relates to the site. Put a chair in the space, sit down, and imagine looking out to the best view; then you'll know where you'll want a large window.

Make a List of What You Want

One valuable thing I learned during my brief career in architecture school was this: Make a complete list of absolutely everything you might want to do or have in your dwelling. Will you want a desk or a couch? How about chairs or some cushions on the floor? Will you want a place to store all your stuff, or a second exit? Write all these things down, then also think about how many people you might want in your place and how much room you'll need for them. For example, the Classic Design described in Part 2 is 6 × 8 feet, which is enough room to comfortably hold a desk or a small table, a bookshelf, and a couple of chairs. Part 3 offers details on building a larger, permanent structure.

Also think about what kind of door you'd like to have and where it will be located. How much light and air do you want to let in? What kind of windows? Some people like their spaces dark and denlike, while others enjoy a lot of light so they can read or see the view outside. A hinged window or two could provide great ventilation on hot days.

Think small, and plan to build just one room at first. You can always add more rooms later (see Chapter 6). But most important: Have fun with it!

PERMITS AND PERMISSION

This is a good time to check with the local "powers that be" about getting a permit or simply a green light to build a small shelter on your property. If you live in a community or subdivision that is heavily laden with owners' agreements or deed restrictions, check the rules to see if you can build in your yard. A homeowners' association agreement may prohibit backyard structures, and some towns and homeowners' associations have ordered backyard dwellings removed because they are deemed unsafe, a fire hazard, unsightly, or all three. Many communities don't want to be exposed to risk from lawsuits or insurance claims. Double-check whether any such restrictions may affect the use of your property to avoid future headaches!

Be aware of any city or county zoning and setback laws. Zoning dictates the use of your property, whether single-family residential, multifamily residential, or commercial. A single-family designation may prohibit a separate "dwelling" in your backyard but not a "shed," depending on how you designate it.

A setback is the distance from a property line that your structure can be located. This is usually 20 feet from the street, 10 feet from the back lot line, and 5 feet from the side lot lines, but all municipalities have their own zoning and setback regulations.

Drawing a Plan

Once you have a site, you can sketch out a plan. A plan, no matter how crude, works a lot better than just starting out nailing boards together. A plan is simply a way to think about what you want and how best to get there. (See Make a List of What You Want on page 27.)

You can draw your plans freehand, you can use a ruler, or you can draw them on graph paper. On graph paper, for example, you can pretend each square represents 3 inches, 6 inches, or 1 foot (you decide), and then count the squares. You can also pretend every inch or half-inch on a ruler is 1 foot on your plan. This way, you can figure out all the dimensions of your structure, including the size of your windows and door. These methods are called drawing to scale, which I'll describe more thoroughly in this chapter and in Part 3.

A floor plan is the easiest to draw. This shows you the floor and walls of your structure and the locations of windows and doors. To get started, use a pencil to draw the walls as double lines. You can erase out spaces where you want your doors and make single lines where the windows will go. Change it around as many times as you want.

Start with a simple plan.

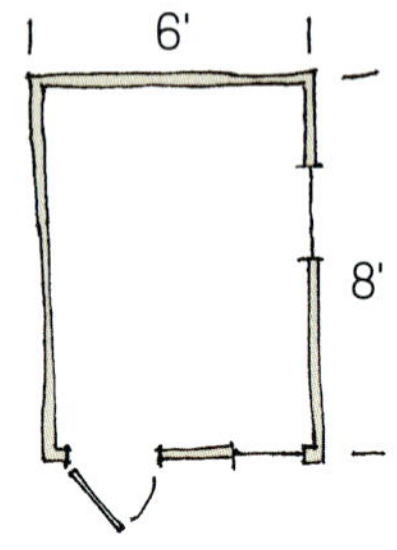

An inside wall will make a cozy nook.

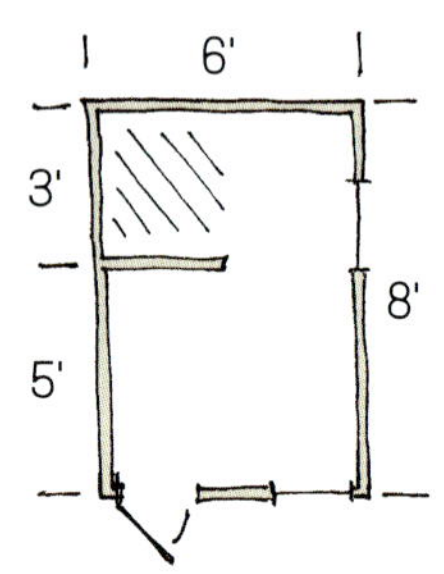

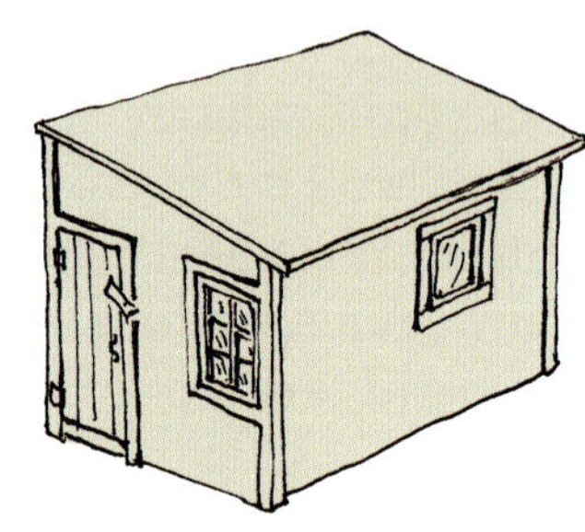

Add rooms if you need to.

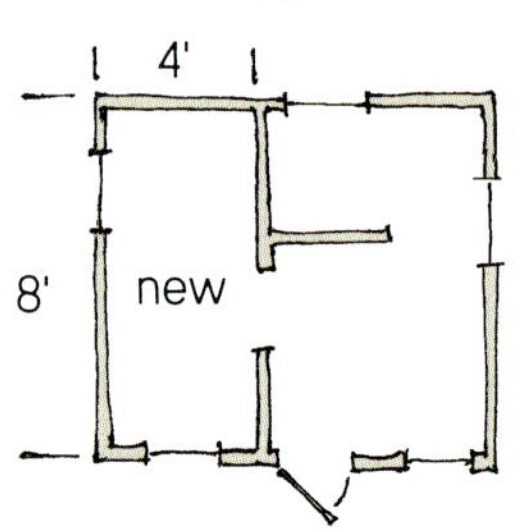

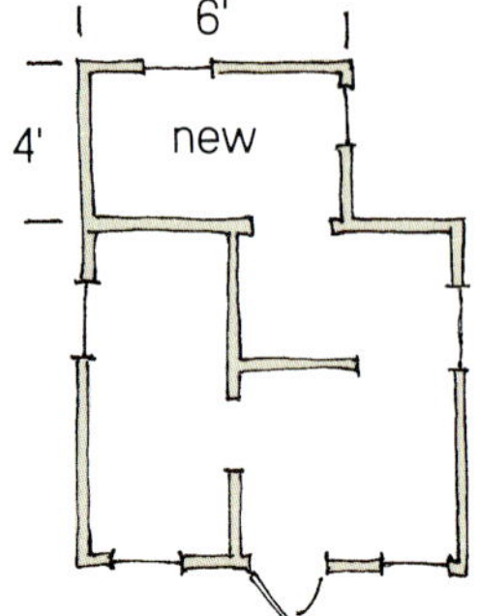

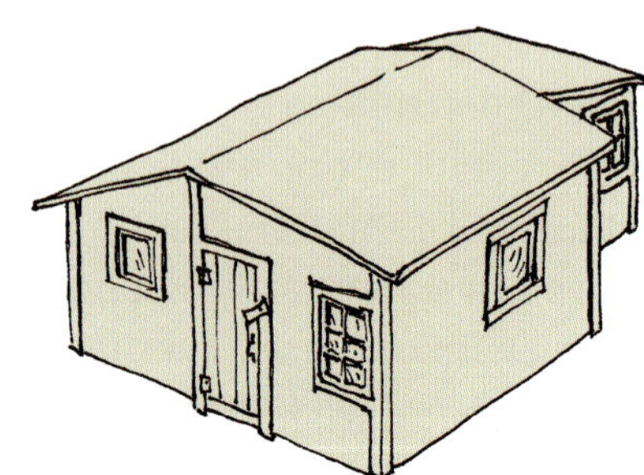

A Simple Floor Plan

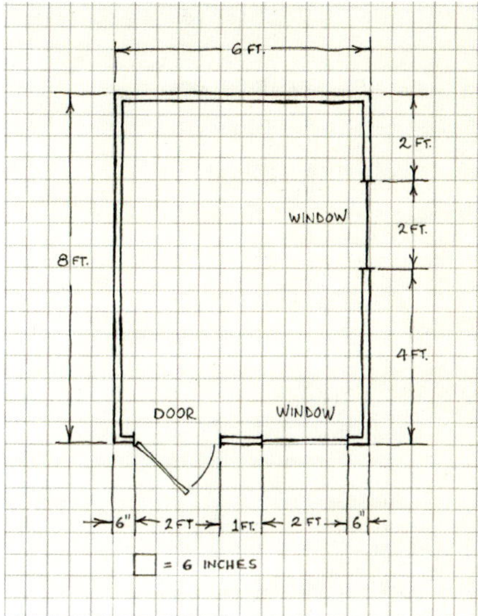

A Simple Elevation

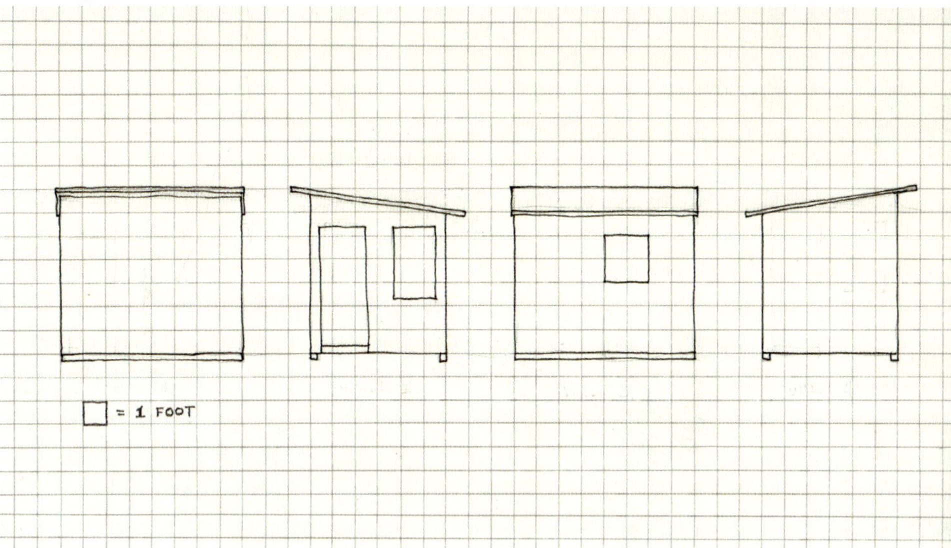

If you want to plan what your walls might look like, how your roof will slant, and where your windows will be, draw an elevation, or side view. This time, imagine how high off the floor the windows should be and which way the roof should slope. (Think about where you will want rainwater to go.) Draw an elevation for each wall, viewed from the outside. Again, you can draw it to scale using a ruler or graph paper.

By drawing plans, you can see how your structure will look, how big it will be, and what will go where. Remember that plans are just ideas, so you can easily change them. This is the best time to make those design adjustments.

Wood, Nails, and Carpenter Language

It's good to know what you'll need for your design so you can avoid collecting a lot of extra stuff. Carpenters and lumberyards have their own words for the wood and other materials they use or sell. For example, a piece of framing lumber can be called a joist, a stud, a sill, a rafter, a trimmer, or a plate, depending mostly on where it is used. Nails are described by all sorts of names, such as roofing, drywall, duplex, casing, galvanized, coated sinker, box, or common, depending on their shape and where they are used. It's kind of like learning a new language, but you'll quickly catch on. These terms (and more) are defined in the glossary at the end of this book for easy reference.

The rest of this chapter is devoted to helping you get familiar with the right materials and how much of them you will need. If you decide to build the Classic Design (described in Part 2), use the materials list on page 41. If you want to build some other size or style of structure, draw your own diagram (see Chapter 8), then count the pieces of wood needed for the floor, the walls, and the roof. There are other estimating methods, but we'll keep it simple for now. As you count the pieces, make up

Carpenter Language

These are the names, in carpenter language, of the pieces of wood that will make up your structure. It's an easy language to get used to.

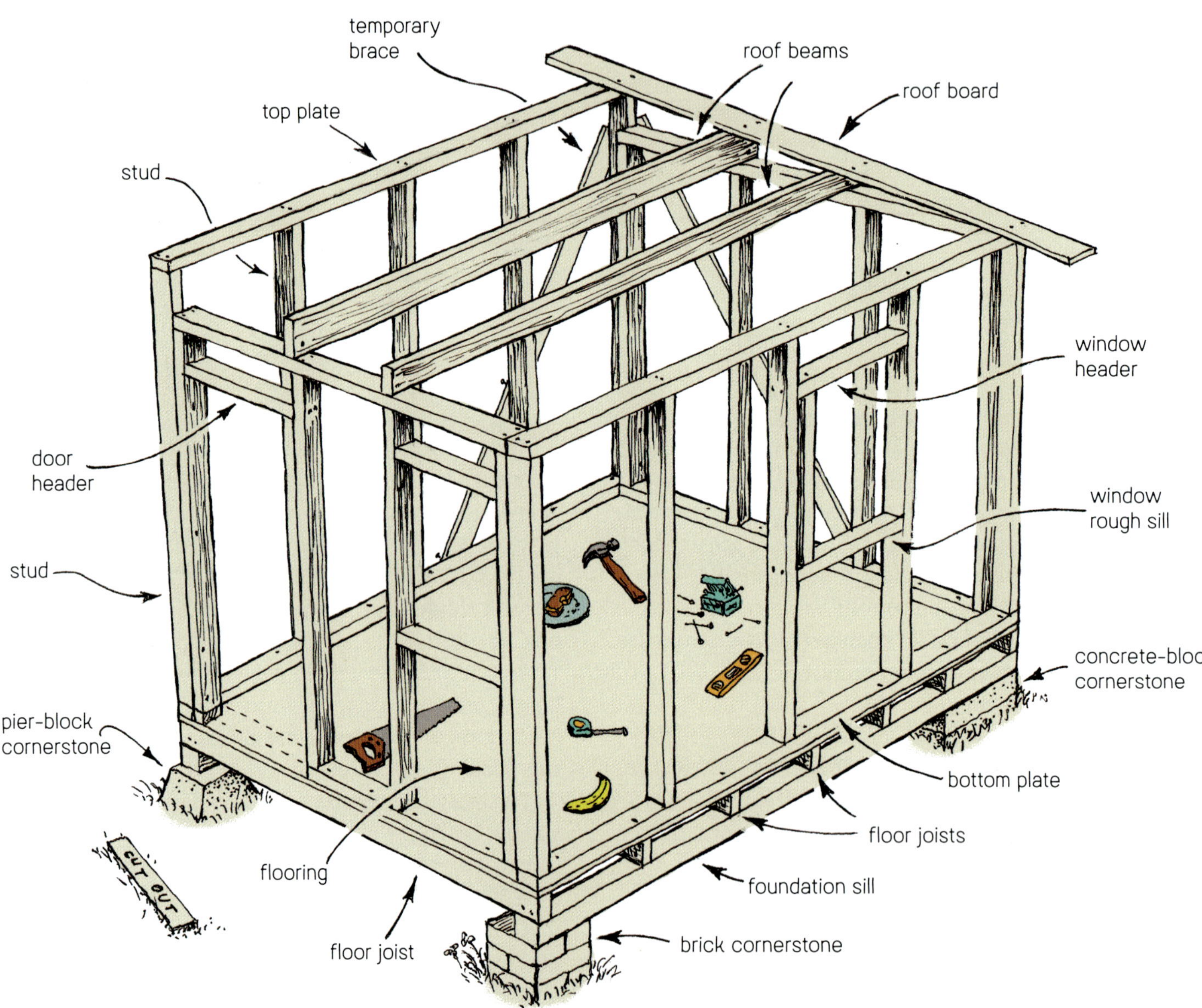

a materials list like the one for the Classic Design. The list is just an estimate, so you might have some extra pieces or you might have to get more supplies, especially if your plans change (which they often do!).

Note: If you buy too many boards, lumberyards are generally nice about returns if the materials are in good condition and you bring your sales receipt with you.

Finding Materials

Before you begin, consider the pros and cons of new versus used building materials. The easiest way to build your shelter is to use all new materials. Modern windows and doors are usually more energy efficient and fairly easy to install, but the cost of new units can add up fast. You can save a lot of money—and perhaps add a unique look to your creation—by collecting used doors, windows, and fixtures instead. This will have to be your call.

I encourage you to use recycled materials as much as you can. (New materials are discussed in the following pages.) Ask family, friends, and neighbors if they have any unwanted boards, windows, doors, bricks, shingles, roofing, siding, hinges, wallpaper, carpet, or paint. Cruise the streets and alleys in your neighborhood for cast-off materials on trash day. Salvaging the materials for your build can be a challenge, but it can be done.

Check out architectural salvage yards, second-hand shops, or Habitat ReStores near you. (See Barn Windows and Other Salvage on page 32.) Finding recycled options lets your imagination take off!

It's best to have your doors and windows (either new or used) on hand to make sure your walls' openings are as accurate as possible to fit them. The openings you build in your rough framing will need to be large enough to fit not only the door or window but also the framework around it.

FOUNDATION

For the structure's foundation, you'll need four cornerstones. These can be anything stonelike, such as large flat stones, concrete blocks, pavers, pier blocks, or bricks. Concrete retaining-wall blocks also work well. (For setting pier blocks in concrete or using concrete tube forms for an even stronger foundation, see Concrete Footings for Uneven Ground, High Winds, Earthquakes, and the Like on page 113.)

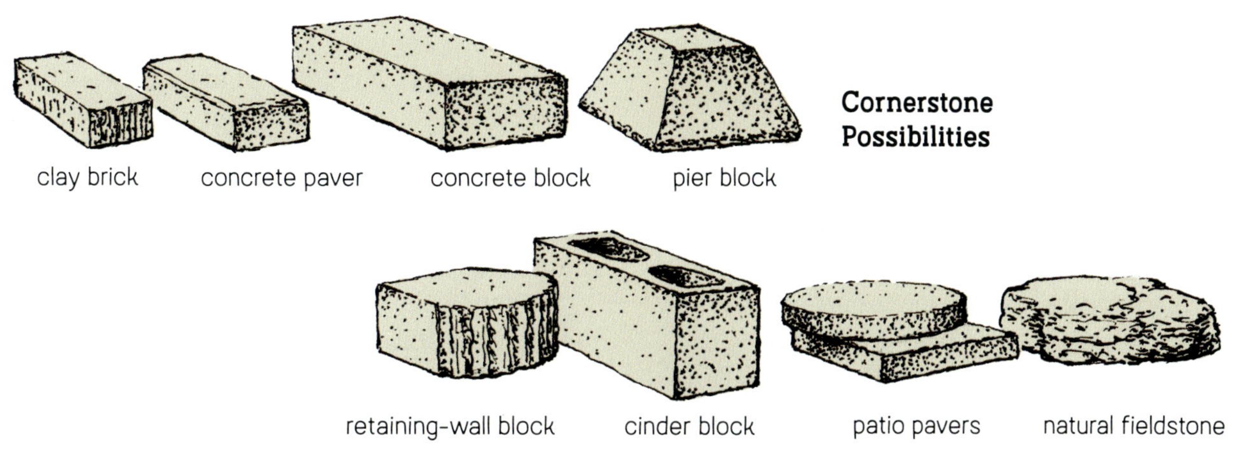

Cornerstone Possibilities

Barn Windows and Other Salvage

If you can't find suitable recycled windows, some of the big lumberyards sell barn sash windows. These inexpensive four-pane or six-pane wood-framed windows are perfect for barns, sheds, and other backyard structures.

You can also find recycled windows, doors, and boards at used building materials outlets. The nonprofit Build Reuse (buildreuse.org) lists stores and salvage yards that sell used building materials in every state in the US. You can also search the internet for local options under "Building Materials—Used."

Habitat for Humanity operates Habitat ReStore, a nationwide chain of recycled-materials stores. You might find one near you at habitat.org/restores. Goodwill and St. Vincent de Paul stores often sell inexpensive windows, doors, paint, carpet pieces, wallpaper, shelves, and other goodies with which to finish or furnish your structure. Free or low-cost wooden pallets can often be found at construction sites, home improvement stores, grocery stores, and schools (ask before taking them!). Also, hardware store paint departments sell returned custom-mixed paint at low prices. Just ask if they have any "returned custom-color" paint.

EVALUATING QUALITY OF WOOD

It's okay to be picky and only select the best boards out of the pile for your build. When choosing, here's what to watch out for.

- **Bent or warped boards.** Look down the length of the board as if you were aiming it like an arrow to see if it's curved or bent. You'll find that most boards are warped a little, so only put aside the ones that are *really* bent.
- **Knots and cracks.** If there are big knots or knotholes (more than half the width of the board) or big cracks (called *checks*), put the board aside.
- **Pitch.** If you see yellowish, sticky goo on a board, it is likely pine pitch. It's nearly impossible to get this stuff to come off your hands or anything else, so set this board aside.

If you're at a lumberyard or home center and there aren't enough good boards to pick through, you can always ask a salesperson to cut open one of the units, or banded stacks, behind the loose boards. Remember, it's their job to be nice to you! Once you've pulled out all the boards you want, be sure to restack the ones you've rejected.

FLOOR, WALL, AND ROOF FRAMING

For the skeleton or frame of your structure, you'll need 2×4 boards. When they hold up your floor, they are called joists. If they are part of the walls, they are called studs.

At the big lumberyards and home centers, the framing lumber—such as 2×4s and 2×6s—is usually stacked in big piles on the floor. Every pile or stack has a sign describing the name of the board, such as "stud" or "standard board," the dimensions, and the price per piece of wood.

Studs, boards, and other lumber are usually labeled with a grade and the kind of tree they came from, such as hemlock, fir, or pine. Look for lumber that is graded no. 2 or better or that is labeled "stud grade." The lumber should specify that it has been kiln-dried; kiln-dried lumber won't shrink or warp as much as "green" or undried lumber. Most of the big lumberyards sell kiln-dried no. 2 and better framing lumber.

For our design's frame, 6-foot-long 2×4s are the cheapest and are long enough for most of the studs. If you can't find 6-footers, then buy 7- or 8-foot studs instead.

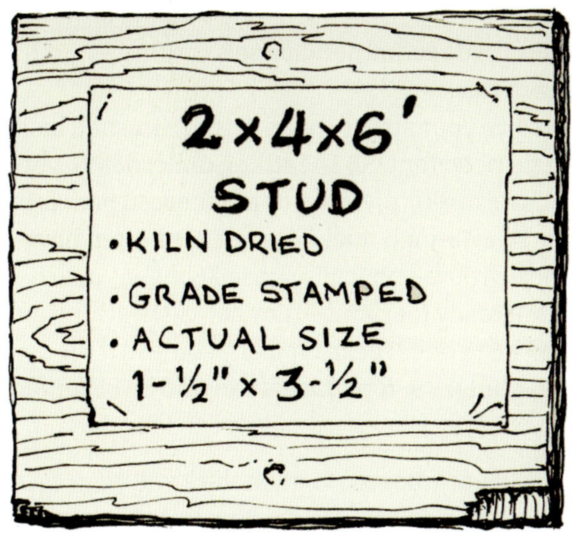

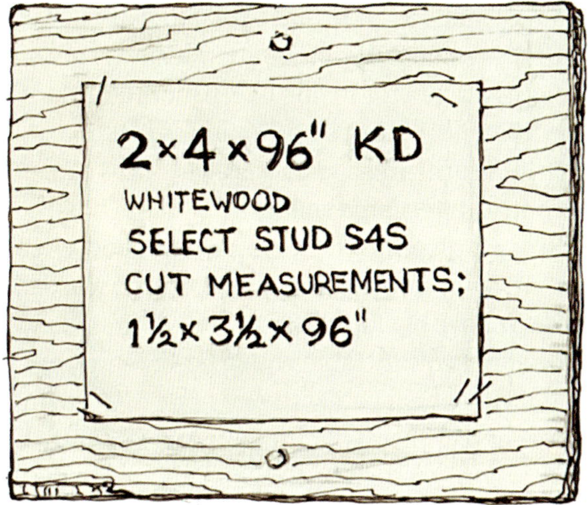

These lumberyard signs tell you that this wood was dried in an oven or kiln, and the grade is stamped or printed on each board. The signs also tell you the exact size of the boards as well as the price. "S4S" means that the board was planed or surfaced on all four sides. The term "whitewood" means any kind of pine or hemlock, and "KD" means "kiln-dried."

Some lumber is labeled "pressure-treated." This lumber has been treated with toxic chemicals to make it rot-resistant. You wouldn't want to use it in your space except for your foundation sills, which are close to the ground and subject to moisture and insects. When you saw pressure-treated wood, wear a dust mask. Ordinary untreated wood will work fine for the rest of your construction, as long as it's away from the soil.

One more thing: A big lumberyard often has a bargain area with odd lots or damaged piles of wood for sale at about half the normal cost. This material is usually special-ordered, slightly weathered, banged up, or otherwise unfit to be sold at the regular price. The bargain area might contain some fence boards, abnormal roofing, odd-colored decking, a whole pile of 1×4s for sale as a unit, or who knows what. If any of this stuff fits your lists, snap it up! Most lumberyards also sell short lumber pieces or scraps up to 4 feet long. These are usually displayed in bins labeled "value wood." You can always use these shorter boards to fill small spaces while covering your walls, so pick through them to find some good deals.

Using 2×3s

Though 2×4s are most common and can sometimes be found (recycled) for free, if you have to buy lumber, you might think about using 2×3s to frame the walls. They are usually cheaper than 2×4s and are plenty strong enough. If you use 2×3s, remember that they are 1 inch narrower (but just as thick) as 2×4s, which will affect the measurements of your walls. Don't use 2×3s for floors, though; they aren't strong enough.

SHEATHING

For the outside skin of your build, you can use common boards, sheets of plywood, oriented strand board (OSB), or other thin lumber called sheathing.

"Common" or "standard" boards are 1×4, 1×6, or wider pine boards with some knots and other minor defects. "Premium," "select," or "clear" pine boards look nicer but are far more expensive. Lumberyards also sell boards called car siding, which are 1×6 or 1×8 pine boards that have tongue-and-groove edges (see the illustration below). These look attractive and are strong and cheap. Boards are easier to saw and put up one by one than the large plywood sheets. New boards can be more expensive, but remember that old fence boards, barn boards, shelf boards, or anything ¾-inch thick will work for sheathing (see page 63).

If you have a power saw and someone to help you, you can use OSB, the cheapest panel sheathing. It is made of strands or chips of wood that are glued together under great pressure. It comes in ⅜-inch, 7⁄16-inch, or ½-inch thicknesses and is meant to underlie the finished siding.

You can also use CDX plywood for sheathing. Plywood is made up of plies, which are thin layers of wood laid crosswise over each other and then glued together under pressure. CDX plywood has a fairly smooth C-grade side, a rougher D-grade side (with knotholes or cracks), and an exterior glue so it won't fall apart in the rain. OSB and plywood normally come in 4 × 8-foot sheets. For a backyard dwelling, I'd recommend panels that are ⅜-inch or 7⁄16-inch thick.

To give your structure an instant finished look with no need for OSB sheathing underneath, you can buy exterior plywood or engineered panels in ⅜-inch or ⅝-inch thicknesses. These often have vertical grooves typically spaced 8 inches apart and are ready for paint—they act as sheathing and siding all in one. These typically come in 4 × 8-foot panels, but 4 × 9-foot panels can also be ordered.

FLOORING

Generally, any boards ¾-inch thick will work fine for flooring. You can use decking boards, so called because they're used for backyard decks; they are usually 5½ inches wide and a full 1 inch thick. Look for untreated decking, since chemically treated lumber is unsafe for inside use. In addition, some varieties of decking are now made of plastic, requiring a drill and special screws.

Some lumberyards sell "ranch-grade" boards, made of rough pine. These might be a good, cheap alternative for flooring if they are not too warped and don't have big knotholes.

If you are comfortable with a power saw—or someone can help you with one—buy ¾-inch-thick tongue-and-groove OSB for your floor. This stuff is less expensive than boards and is very strong. It's what builders use for house floors. It comes in 4 × 8-foot sheets, which may be more than you need, but you can use the extra on a wall or the roof.

Common Boards

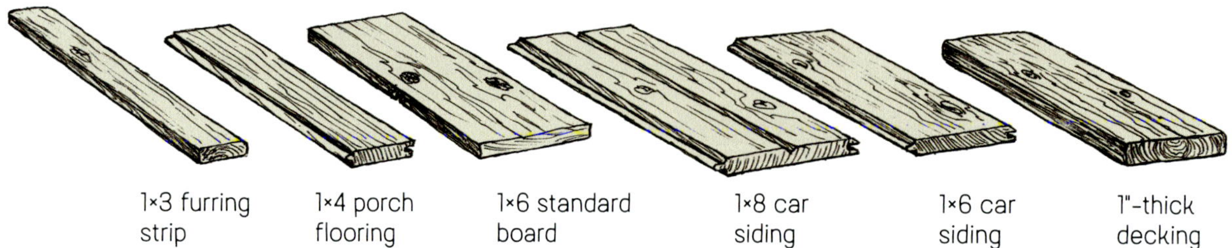

ROOFING

A simple roof has two parts: the boards that make it strong, and the roofing or roof covering that keeps the rain out.

For the boards, you can use ¾-inch-thick standard boards, car siding, plywood, or OSB. Be sure to use ⅝-inch- or ¾-inch-thick sheets or boards, because the roof beams are spaced too far apart for the thinner sheets. For its price, strength, and good looks inside your shelter, I recommend car siding.

Many roof coverings are available, and since you're building a small structure, you may be able to find enough leftovers from another building project. Ask your neighbors or local contractors. Lumberyards also might have odd piles of roof covering in their bargain areas. Acceptable options are roll roofing, asphalt or fiberglass roof shingles, metal sheets, or just about anything else that will shed water and not rot. (Note that if you use metal roofing, you'll need metalworking tools to cut it and fasten it in place.) If you have to buy roofing, look for either 30-pound felt, which is a felt fabric soaked in tar (also called tar paper), or mineral-coated roll roofing, which is a thicker felt tar paper covered with fine, colored gravel, and which lasts much longer.

NAILS AND SCREWS

When you look for fasteners (nails, screws, and bolts) at a hardware store, you'll find rows and rows of them displayed in little boxes. The boxes have labels describing the kinds of nails they hold and their size, such as 4d, 8d, 16d, etc. These sizes refer to their length and are based on the old English pennyweight system. For example, a 10d, or tenpenny nail, used to cost 10 pence per hundred. It is 3 inches long. You'll be using mostly 6d to 16d nails.

Roll Roofing

black, white, or colored granules; lasts 10 to 15 years

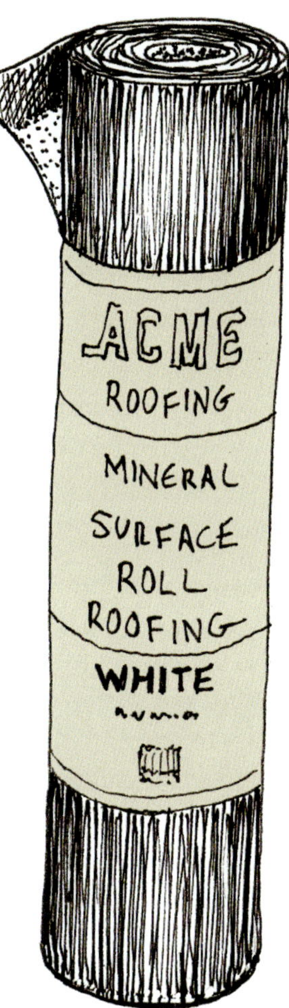

all black (no granules); lasts 2 to 3 years

For a longer-lasting roof, select "mineral surface/90-pound" roll roofing. One 36-foot-long roll will cover your roof. Do not confuse it with 15- or 30-pound roll roofing, which is meant for shingle underlayment.

Nails and Screws

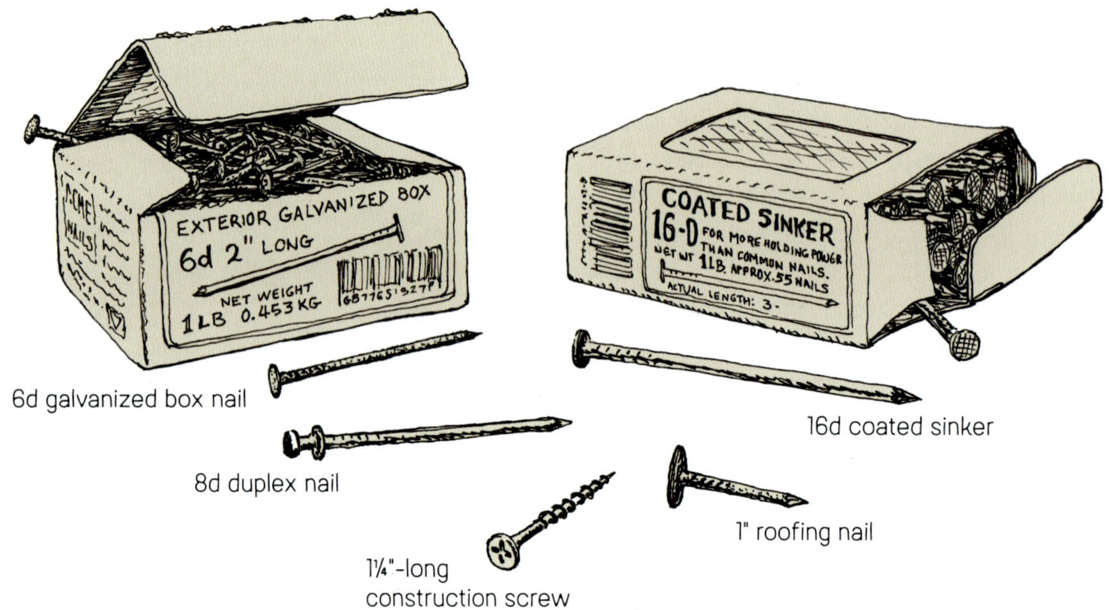

Here are the kinds of fasteners you'll most likely need.

- **Coated sinkers.** Coated sinkers are the cheapest nails to buy and the easiest to pound in. Sinkers are coated with a greenish plastic film that makes them easy to sink into the wood. The same coating also sticks to the wood once the nails are in, giving the nails great holding power. Most builders in the US use them. You will want 12d or 16d (your choice) coated sinkers to nail the 2×4s together, and 6d or 8d (your choice) coated sinkers to nail the sheathing to the 2×4s.
- **Duplex nails.** These nails are used for temporary bracing. They have two heads so that when you no longer need the bracing you can easily pull them out. For starters, buy a 1-pound box of 6d.
- **Galvanized box nails.** Because they're galvanized, these nails won't rust, and they're best for your exterior trim boards. Get a 1-pound box of 6d.
- **Galvanized roofing nails.** For the roof, buy a 1-pound box of 1-inch-long galvanized roofing nails. These have extra-wide heads on them to hold down tar paper, roll roofing, or roof shingles.
- **Construction screws** (optional). These come in many sizes for different applications, such as installing paneling, drywall, and decking. They are sold in 1-pound or 5-pound boxes (or larger). An economical type for a door would be galvanized exterior Phillips-head screws. Get a 1-pound box of 1¼-inch-long construction screws if you are going to build your own door; they also work well for attaching hinges.

HINGES

You'll need hinges for your door. If you're reusing a door, use the original hinges that came with it. If it has no hinges, ordinary door hinges will work fine. If you have to buy new hinges, get a set of two T hinges, which are used for gates and shed doors.

Okay, now you know enough about building materials to make a materials list and head to the lumberyard. You might want to use the materials list for the Classic Design (see page 41) if your plan is similar. Don't hesitate to use the carpentry words we've discussed when asking for wood at the lumberyard. The sales staff will appreciate it.

Hinges

All use ¾"- to 1"-long flathead wood screws.

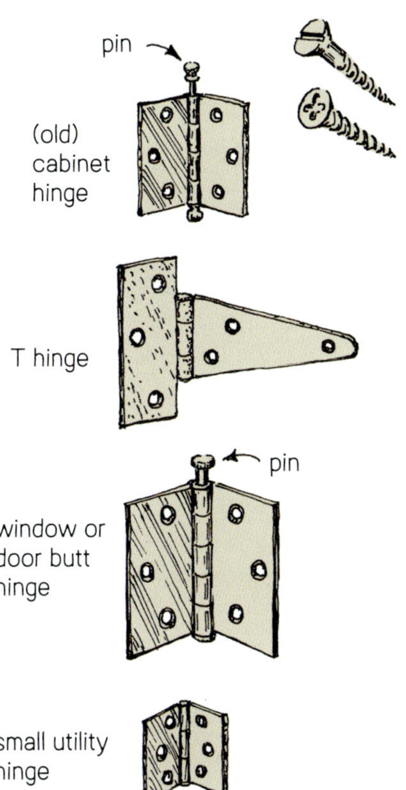

A Note to Parents About Playhouses

Building a playhouse or clubhouse with your children (or letting them build it) is a great way to spend time together, as well as being a powerful confidence-building experience. You'll show them that you care about them and you trust them not to do stupid things with sharp tools. If you have never built anything yourself, no problem! You can all be kids learning hands-on skills together while gaining confidence in your ability to create something substantial.

Kids love it when you listen to them, so start by asking them what they would like if they had a place of their own to play in, using the same questions you'd ask yourself (see page 27). Let them draw some plans or write up a list of features.

If your kids are younger, say 8 to 10 years of age, offer to help design and build their project, but give them as much space as you can. If they are 11 or older (or think they can do it all themselves), try not to impose your presence any more than they request.

If you help them with their building desires and plans, try to steer them to a reasonably sized structure. Suggest secret nooks, hidden exits, or cozy alcoves; kids love small, low-ceilinged places to curl up in. On the other hand, kids do grow up, so allow room for a taller addition later on. Don't forget windows for light and ventilation.

Let the kids "improve" their hideaway as much as they want to. Give them room to experiment with wild paint-color combinations or graffiti, odd-shaped additions, and gardens if they so desire. It may look like an eyesore at times, but if you limit the rules to tidying the yard and putting away the tools when they are finished, they will stay engaged with their playhouse or clubhouse, possibly for years—and you will know where they are!

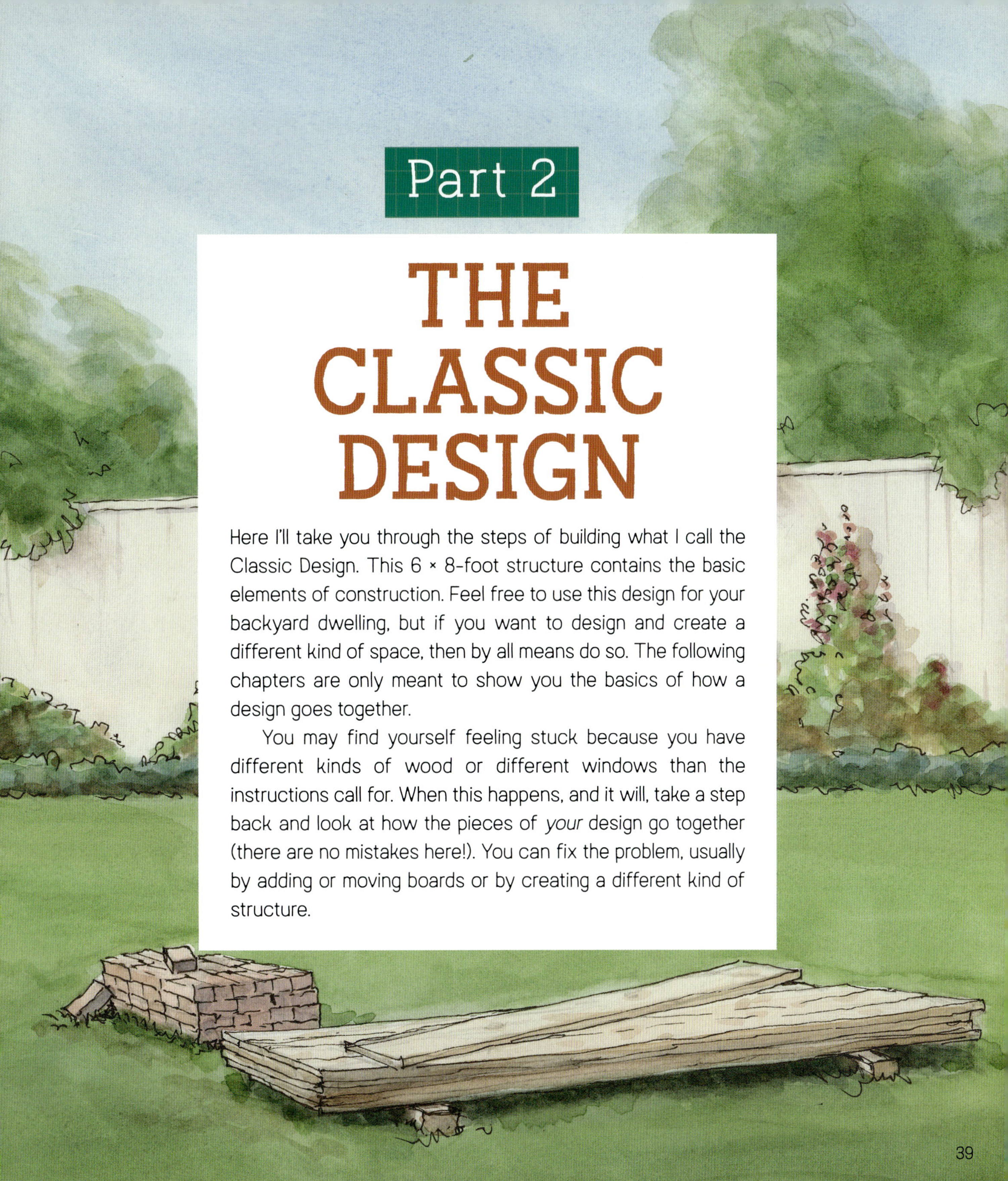

Part 2

THE CLASSIC DESIGN

Here I'll take you through the steps of building what I call the Classic Design. This 6 × 8-foot structure contains the basic elements of construction. Feel free to use this design for your backyard dwelling, but if you want to design and create a different kind of space, then by all means do so. The following chapters are only meant to show you the basics of how a design goes together.

You may find yourself feeling stuck because you have different kinds of wood or different windows than the instructions call for. When this happens, and it will, take a step back and look at how the pieces of *your* design go together (there are no mistakes here!). You can fix the problem, usually by adding or moving boards or by creating a different kind of structure.

Materials for Building the Classic Design

The materials on the list on the facing page will allow you to build a 6 × 8-foot structure. Each of the chapters that follow includes its own part of this list, so you can acquire the supplies for each step of the process as you get there. Of course, if you like, you can copy this page, take it with you to the lumberyard, and come back with all you need for the entire construction. But don't forget about the recycled wood, windows, and other materials you might find along the way as well.

When items include multiple options, I recommend the first item listed.

Floor Plan for the Classic Design

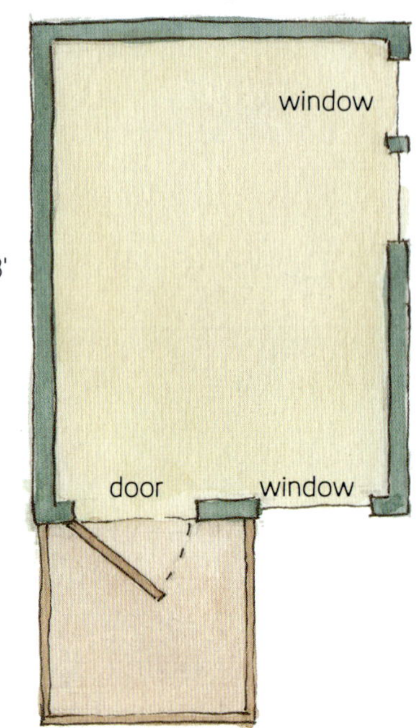

CLASSIC DESIGN COMPLETE MATERIALS LIST

PART	QUANTITY	DESCRIPTION
FOUNDATION	4	pier blocks, concrete blocks, or large flat stones; *or*
	24	bricks
FOUNDATION STAKES	4	1×2 or 2×2 scraps, about 2 feet long
FOUNDATION SILLS	2	4×4s, 8 feet long, pressure-treated
FLOOR JOISTS	7	2×4s, 6 feet long
FLOORBOARDS	14	1×6 car siding or pine boards, 8 feet long; *or*
	1½	4 × 8-foot sheets ⅝-inch- or ¾-inch-thick plywood; *or*
	1½	4 × 8-foot sheets ¾-inch-thick tongue-and-groove OSB
WALL STUDS	20	2×4s, 6 feet long
WALL PLATES	4	2×4s, 8 feet long
	4	2×4s, 6 feet long
UPPER ROOF BEAM	1	2×6, 8 feet long
LOWER ROOF BEAM	1	2×4, 8 feet long
BRACING	8	1×4 furring strips or boards, 8 feet long
WALL SHEATHING	65	1×6 car siding or pine boards, 8 feet long; *or*
	5	4 × 8-foot sheets 7⁄16-inch- or ½-inch-thick CDX plywood; *or*
	5	4 × 8-foot sheets 7⁄16-inch- or ½-inch-thick OSB
ROOF BOARDS	20	1×6 car siding or pine boards, 8 feet long; *or*
	14	1×8 car siding or pine boards, 8 feet long
ROOF COVERING	1	roll mineral-coated roofing (better) or 30-pound felt underlayment (okay)
TRIM BOARDS	2	1×3s or 1×4s, 10 feet long, for roof edging
TRIM BOARDS	14	1×3s or 1×4s, 8 feet long, for corners, doors, and windows
WINDOWS	2	20 × 25-inch barn sashes, or other size, depending on what you find
WINDOW HINGES	4	small utility or old cabinet hinges, with screws
DOOR	1	recycled or salvaged closet or other small door, or build one yourself (see page 66)
DOOR HINGES	2	6- or 8-inch-long T hinges, or 3 × 3-inch butt hinges, with screws
NAILS	2 pounds of each	16d coated sinkers; 12d coated sinkers; 8d coated sinkers; 6d coated sinkers
NAILS	1 pound of each	6d galvanized box nails, for the trim boards; 6d duplex nails, for the brace boards; ¾-inch or 1-inch galvanized roofing nails
SCREWS	1 pound	1¼-inch Phillips-head construction screws (optional; useful if you're building your own door)

·CHAPTER 3·

Building the Foundation and Floor

It's time to dig in the dirt, set down a foundation, and build a floor. To start, you'll need your nine essential tools, plus a shovel and the materials listed below. Among the choices, I recommend the first item that is locally available or most affordable.

FOUNDATION AND FLOORING MATERIALS

PART	QUANTITY	DESCRIPTION
FOUNDATION	4	pier blocks, concrete blocks, or large, flat stones; *or*
	24	bricks
FOUNDATION STAKES	4	1×2 or 2×2 scraps, about 2 feet long
FOUNDATION SILLS	2	4×4s, 8 feet long, pressure-treated
FLOOR JOISTS	7	2×4s, 6 feet long
FLOORBOARDS	14	1×6 car siding or pine boards, 8 feet long; *or*
	1½	4 × 8-foot sheets ⅝-inch- or ¾-inch-thick plywood; *or*
	1½	4 × 8-foot sheets ¾-inch-thick tongue-and-groove OSB
NAILS	1 pound of each	6d galvanized box nails, for the trim boards; 6d duplex nails, for the brace boards; ¾-inch or 1-inch galvanized roofing nails

Building the Foundation

The Classic Design is 6 feet wide and 8 feet long. To build its foundation, you'll need to set cornerstones of concrete blocks, bricks, or flat stones at each corner. If you have to buy your cornerstones, pier blocks or concrete blocks work best.

Look at your building site and decide where you want the corners of your structure to be. If the site is cluttered with weeds or junk, clear it out before building. A clean site is a safe site. Start out by finding the foundation corners and pounding in stakes.

SET THE STAKES

Step 1. For your first corner, pound in a stake, which we'll call Stake #1.

Step 2. Hook your tape measure on the stake (or have a helper hold it there) and pull it to the 8-foot (96") mark to find the location of Stake #2, and pound that one in.

Step 3. Now turn and measure 6 feet (72") from Stake #2 to set Stake #3.

Step 4. Now measure from both Stake #3 *and* Stake #1 to set Stake #4.

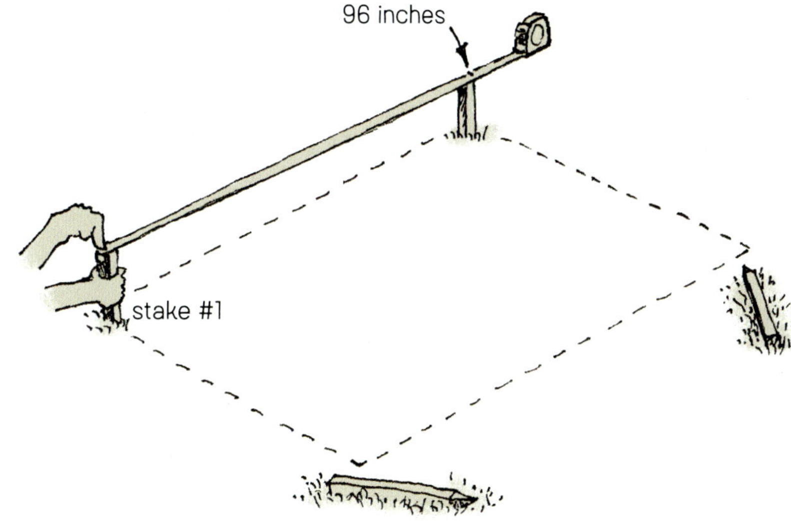

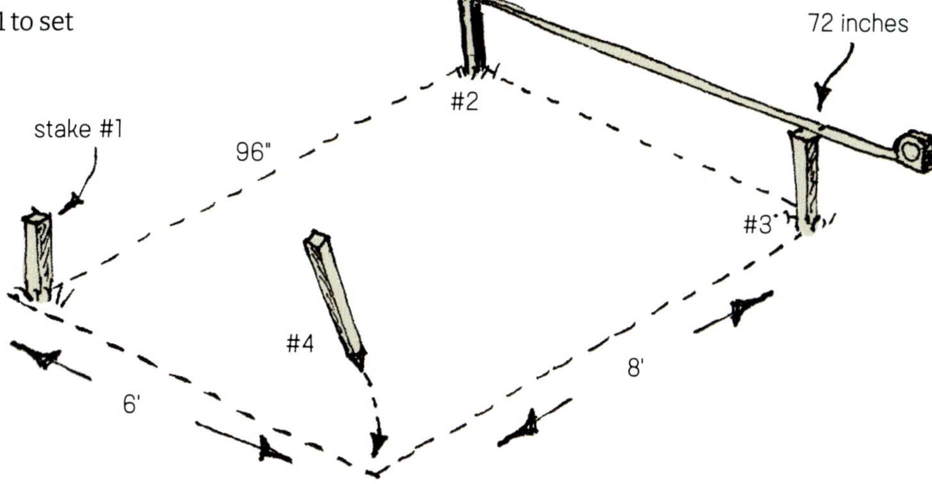

SQUARE THE FOUNDATION

Once the stakes are in, check them to make sure your foundation is square. "Square" means each of the four corners forms a 90-degree angle.

Step 1. Get someone to help you measure diagonally across your stakes (see the drawing at right). For a 6 × 8-foot foundation, the diagonal measurement will be exactly 10 feet.

Step 2. Pull up and reset the stakes until you get this measurement in both directions. This method works best on fairly flat ground, and don't worry if the diagonal measurements are off by up to an inch; you'll test this again when building the floor frame.

SET AND LEVEL THE CORNERSTONES

The stakes now tell you exactly where the outer corners of your foundation stones are to be set.

Step 1. In each corner, dig out any loose dirt and grass for the cornerstone.

Step 2. When you're ready to set the cornerstone in place, take out the stake. Set the stone, adjusting it as necessary to make sure it's stable.

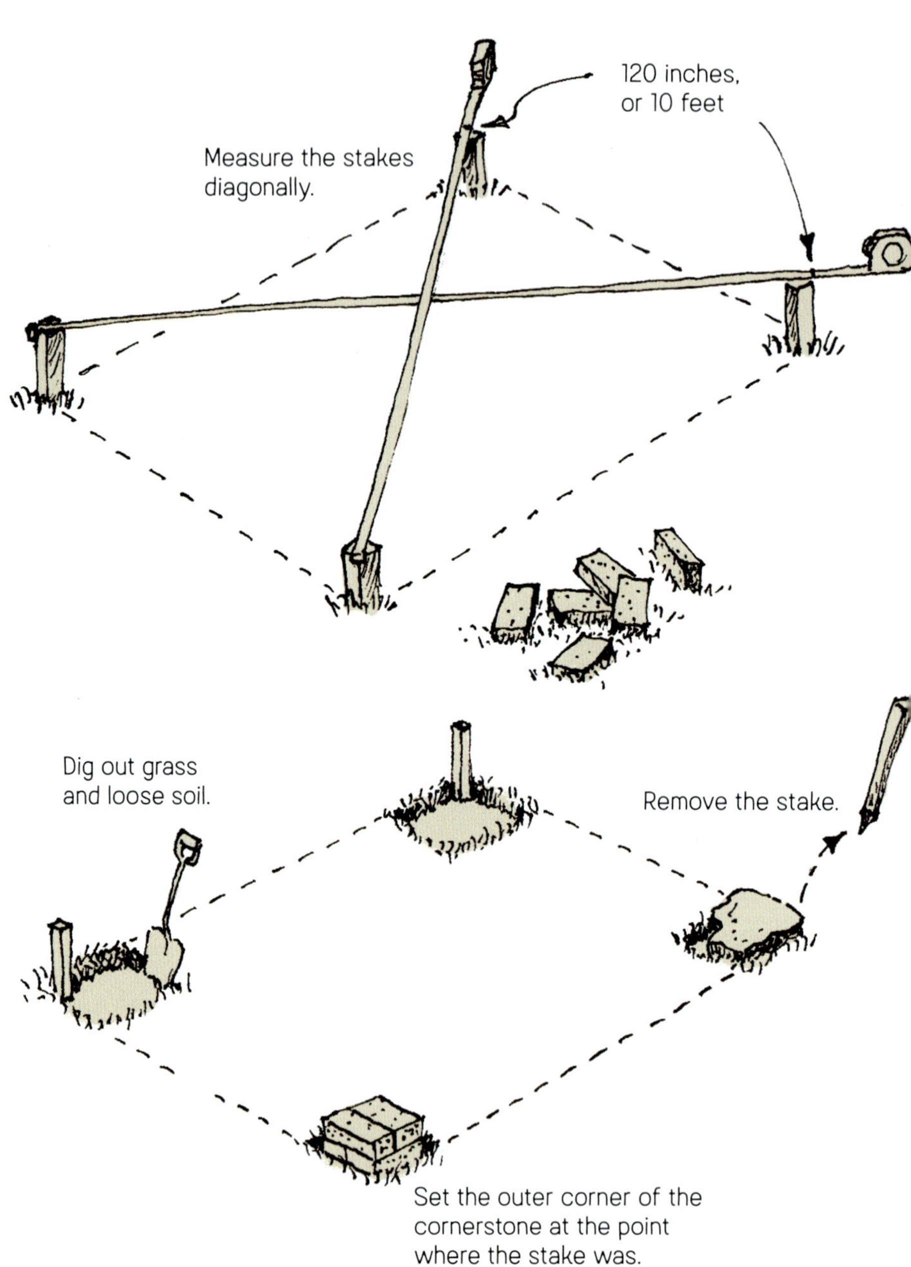

Step 3. Level the cornerstones to ensure that the floor of the structure will be level. You can check for level by laying a long, straight 2×4 across the top of each pair of cornerstones and setting your level on top of it, as shown in the drawing below. If the cornerstones aren't level, you may have to dig out more dirt from beneath one or more, or add bricks or stones. Then check for level again. Be patient; sometimes this takes a while, but starting with a level foundation is key to a successful structure in the end.

A Small Level Plus a Long, Straight Board = A Big Level

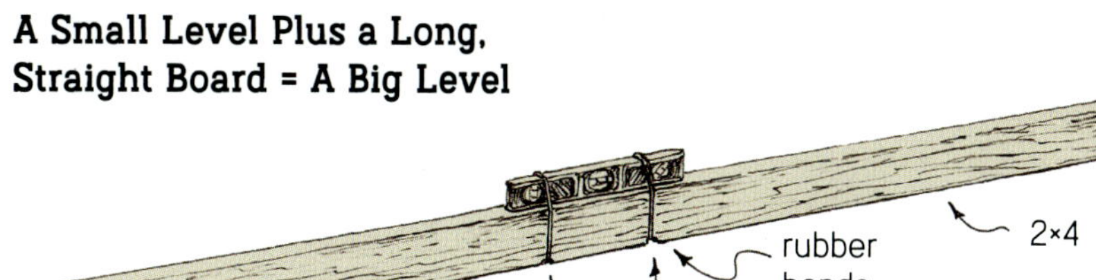

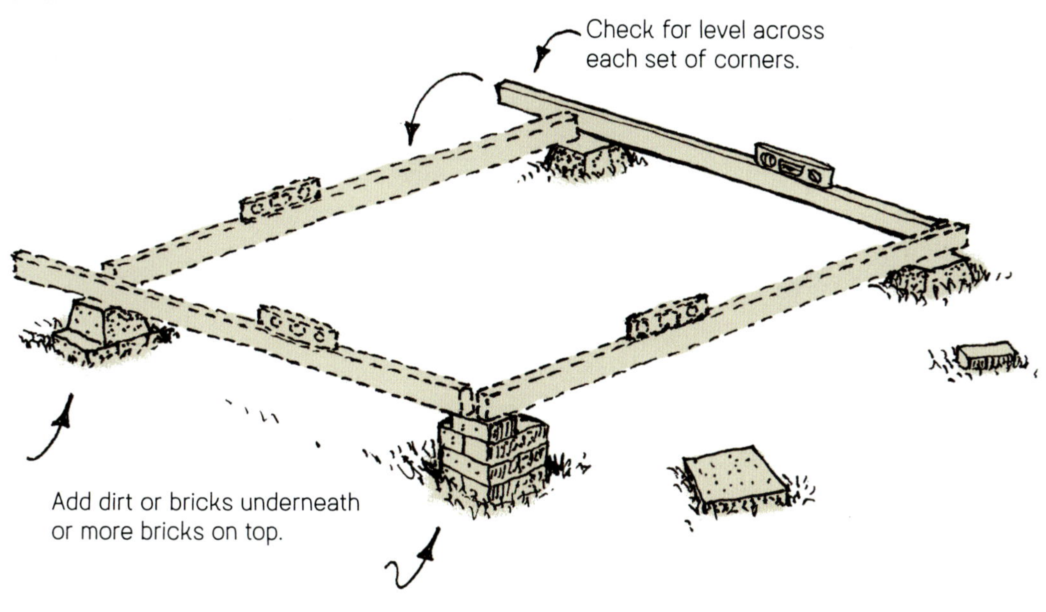

The Floor Frame

Now you'll build a floor, beginning with 4×4 sills, which will support the 2×4 joists; the joists will support the floorboards. The 4×4s are big and heavy, but you'll be able to saw through them with a sharp handsaw.

CUT THE SILLS

Step 1. Set the 4×4 sills on your sawhorses, then get out your tape measure, pencil, and square. Measure to exactly 8 feet (96") along the top of each 4×4 and mark that spot. If the sills are already 8 feet long, or within ¼ inch of 8 feet, you lucked out! If not, use your square and draw a line across each 4×4 at that mark and also down the sides.

Step 2. Begin sawing the first sill. Take your time as you saw steadily through the 4×4. If it wiggles, have a helper hold it down. Use the line you drew as a guide, keeping the saw on the line. Make sure the piece you are cutting off is on the outside of the sawhorse (not between the sawhorses) so it will fall free.

Step 3. Do the same thing for the other sill. Now you're done with the thickest pieces of wood in your entire design.

MARK THE SILLS FOR THE JOISTS

Step 1. Lay both sills side by side on your sawhorses. Hook your tape measure to one end of one of the sills, and with your pencil, make a mark every 16 inches to the other end.

Step 2. Using your square, draw lines over the marks across both sills. These will be the centerlines for your 2×4 floor joists (the boards that hold up the floorboards). This means that the joists will be centered over the lines. Tape measures usually have little arrows at 16, 32, 48, and 64 inches, and every 16 inches thereafter, because this is the standard spacing of joists, studs, and rafters in house building.

Step 3. Once you've drawn all the centerlines, set the sills on the cornerstones. Set your level on each one to check for level, just to be certain.

Marking the Sills

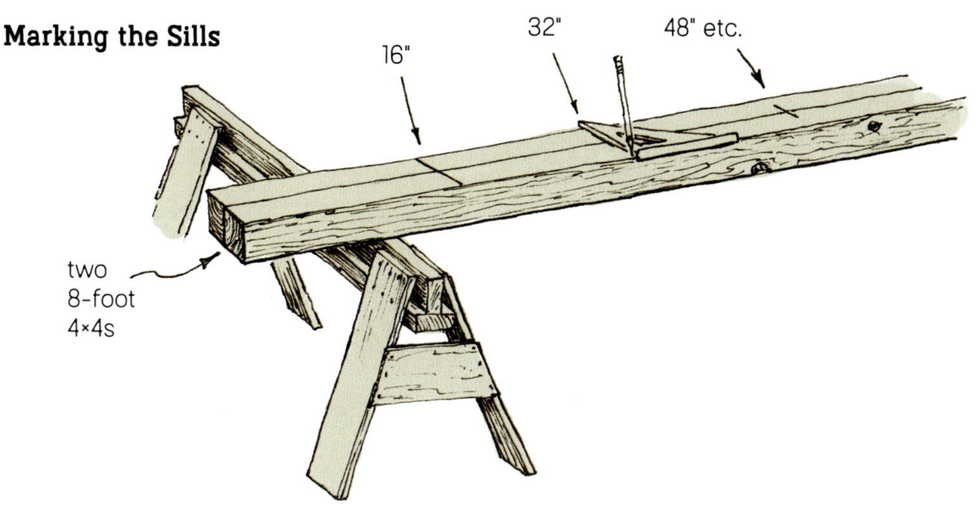

Installing the Sills

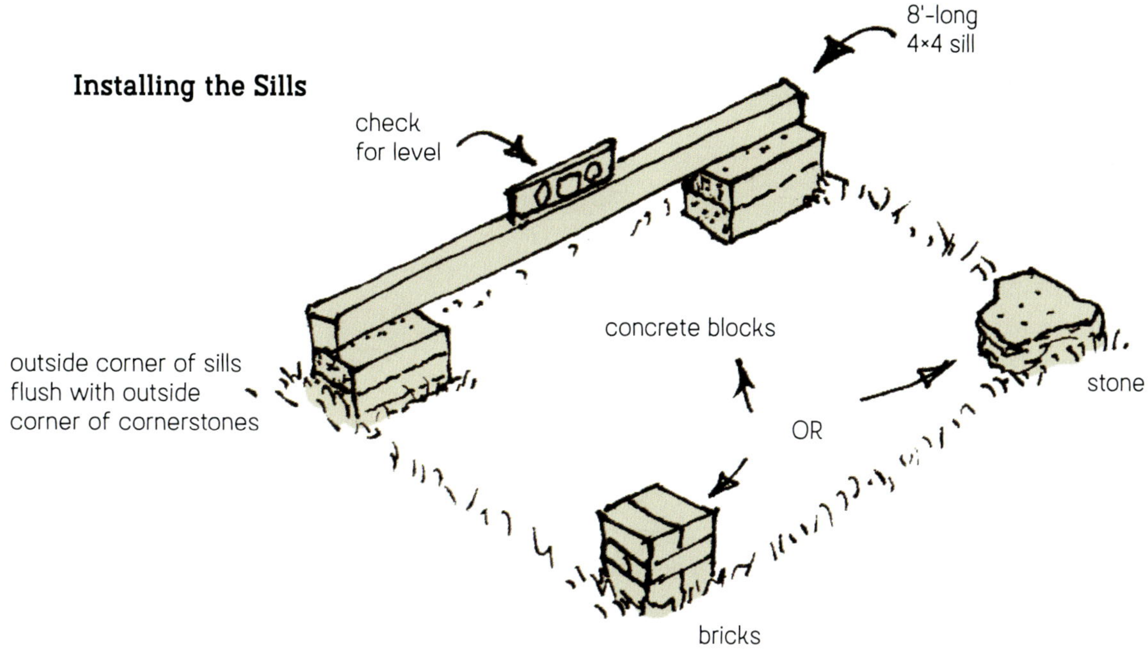

INSTALL THE JOISTS

Step 1. For your floor joists, get seven 2×4s and place them on your sawhorses. Measure and mark each 2×4 at 6 feet long, then draw a cut line using your square. Saw each one to length slowly, watching your cut lines.

Step 2. Set your joists on the sills so they are centered on the centerline marks, then put your level on them for one more test. You might have to add a shingle or thin piece of wood under one or more of the sill ends to finally get the whole thing level.

Step 3. Once the joists are level, nail them to the sills with 6d nails. You'll have to toenail them—that is, drive a nail at an angle through one and into the other. The best way is to hold the joist in place with your knee, tap the nail into the side or end of the joist, near the bottom, and then drive the nail at an angle until it "bites" into the sill. (See the drawing below.) Toenailing can be tricky at first, but it will get easier with practice, I promise.

Step 4. If your floor is crooked, meaning not square and level, it will be a lot harder to build the rest of your structure, so test your floor frame for squareness again by measuring the diagonals. Hook your tape to one corner of your floor, pull it diagonally to the far corner, and see if you get 10 feet, or 120 inches. If it's longer, have a helper hold down or sit on the far corner of the floor frame while you push the opposite (diagonal corner) toward them to lessen the long measurement. Continue until the diagonals are both the same.

Step 5. Test the frame for level one more time, because all the nailing and nudging might have moved things.

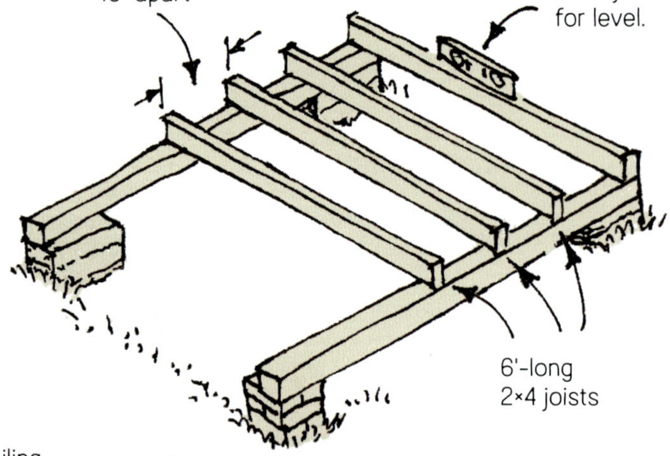

To toenail is to fasten two boards together by driving a nail through one into another at an angle.

Putting Down the Floorboards

Any boards, plywood, or OSB panels, even in combination, will work just fine for flooring so long as all the boards are the same thickness. Car siding, nailed with the grooves facing down, makes a nice, tight floor.

HOW TO INSTALL FLOORBOARDS

Step 1. Set some floorboards on your sawhorses. Measure and cut the boards so they either go across all your joists (8 feet long) or will end over the center of a joist. Since your joists are all 16 inches apart, the floorboards will be 16, 32, 48, 64, 80, or 96 inches long. The longer the better, and every board must end over the middle of a joist.

Step 2. Nail the boards to every joist with 8d nails.

Floorboards

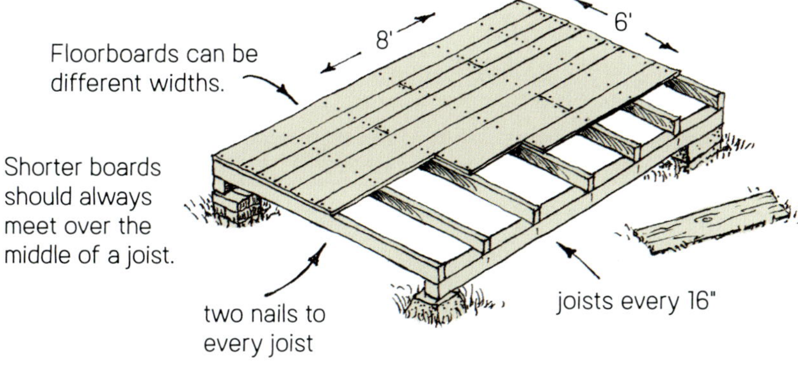

Plywood Panels

one nail every 6" on edges

one nail every 12" inside

Three-quarter-inch-thick tongue-and-groove OSB makes a tight-fitting floor and is easy to install.

joists every 16"

HOW TO INSTALL PLYWOOD OR OSB PANELS

Step 1. If you're using sheets of plywood or OSB, first cut them to size, then set them over the joists. Another method is to set them in place first, then trim off the scrap along the edges with a power saw. Remember to use ⅝-inch- or ¾-inch-thick plywood or OSB. If your floor frame is square, a 4 × 8-foot sheet should fit nicely on your joists, without any cutting. An additional 2 × 8-foot piece will finish the job.

Step 2. Start a nail in each corner of a sheet or panel to set it in place, then pound in a nail every 6 inches along the outside edges and every 12 inches on the inside joists. Use 8d nails.

·CHAPTER 4·

Framing the Walls and Roof

With the foundation and floor complete, you can now build the frame, or skeleton, of your structure. All of the wall frames will have top and bottom 2×4 pieces, called plates, and vertical 2×4 pieces, called studs. Assemble the wall frames on the ground, then set up each frame one by one on the floor. Three of the walls will be 66 inches tall, and the fourth one will be 9 inches taller, or 75 inches tall, so the roof can slope. The roof frame will consist of only two beams, a 2×4 and a 2×6, both 8 feet long.

FRAMING MATERIALS

PART	QUANTITY	DESCRIPTION
WALL PLATES	4	2×4s, 8 feet long
	4	2×4s, 6 feet long
WALL STUDS	20	2×4s, 6 feet long
UPPER ROOF BEAM	1	2×6, 8 feet long
LOWER ROOF BEAM	1	2×4, 8 feet long
BRACING	8	1×4 furring strips or boards, 8 feet long
NAILS	2 pounds of each	16d coated sinkers; 12d coated sinkers; 8d coated sinkers; 6d coated sinkers
NAILS	1 pound	6d duplex nails, for the brace boards

To build your walls and roof frame, you'll need your nine essential tools, a stepladder, and the materials listed above. You'll also need a large, flat spot near your floor on which to put the wall frames together.

Building Wall 1

Let's build the tall wall first. This wall will be 8 feet long, the same length as the floor, and won't have any doors or windows.

FRAME WALL 1

Step 1. First you'll make the plates. Set two 2×4s that are at least 8 feet long on your sawhorses. Measure the 2×4s, and if they're not 8 feet long (or within ¼ inch of it), cut them to length.

Step 2. Put the plates together, on edge, on your sawhorses. Hook your measuring tape to one end, and make marks at 24, 48, and 72 inches. With your pencil and square, draw a centerline across both boards at each mark. The centerline will show where the center of each stud will go. Set the plates aside.

Step 3. For the studs, set out five 2×4s that are 6 feet or longer on your sawhorses. Since wall 1 will be 75 inches high and the thicknesses of the two plates together add up to 3 inches, then the studs will be 72 inches long (75"–3"= 72"), or 6 feet. Measure the 2×4s, and if they're not 6 feet (or within ¼ inch of it), cut them to length.

Step 4. Lay the plates and studs on edge on the ground so that the centers of your studs meet the centerlines on the plates. Using 12d or 16d nails, drive two nails through the plates into the ends of each stud, as shown. As you hammer, hold down each stud with your knee, or step on it, so it will stay put.

Frame Wall 1

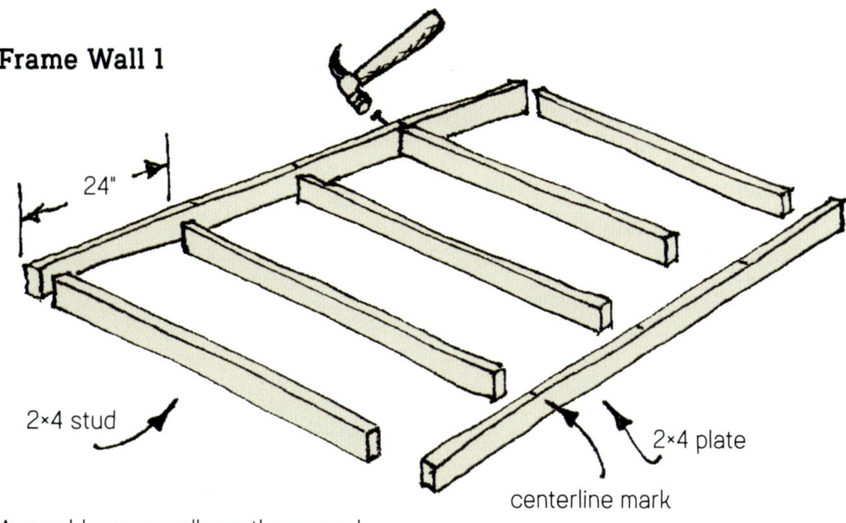

Assemble your walls on the ground, then lift them up on your floor.

INSTALL WALL 1

Step 1. With a helper, carry the wall frame to the floor and set it in place.

Step 2. Nail the wall down by driving one or two 12d nails between each stud through the bottom plate and into the floor. While your helper holds the frame, use your level to check whether the wall is plumb (straight up and down). If it isn't, ask your helper to gently nudge the wall frame until the bubble in your level is in the middle.

Step 3. To keep the wall plumb, nail on temporary brace boards, as shown on the drawing at right. Here it can be helpful to have three people working: one to hold and adjust the wall, one to check for plumb, and one to attach the bracing. You will pry off the brace boards later, so don't nail them in too tight. Use 6d duplex nails, the two-headed nails that are easy to pull out.

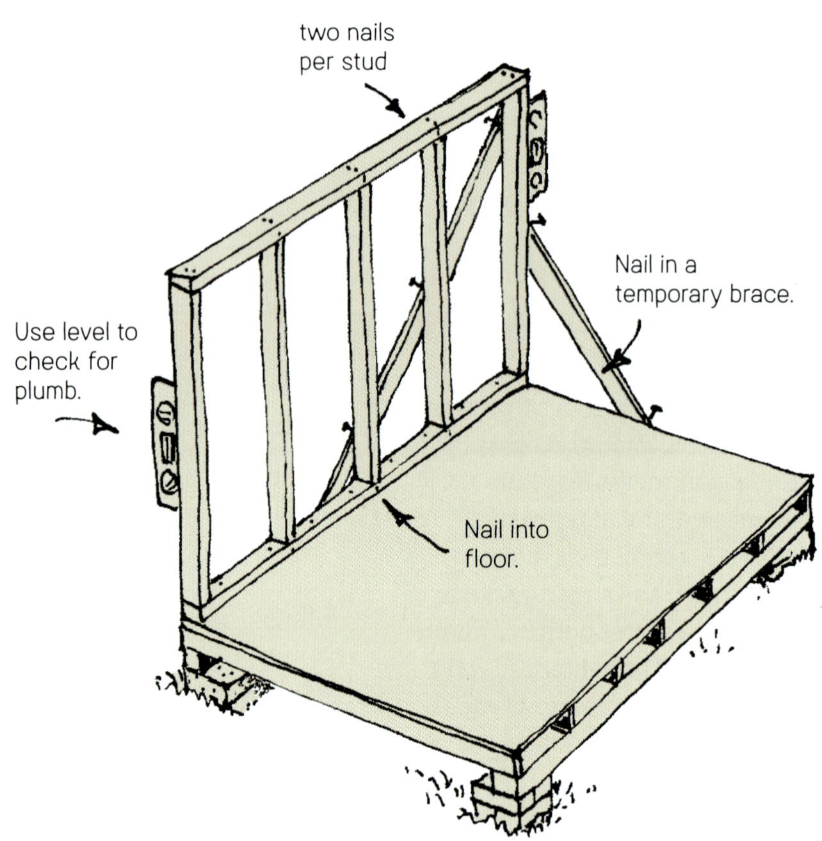

The Right Nails

Always use 10d, 12d, or 16d nails when nailing through 2×4s. Although 6d or 8d nails work well for toenailing 2×4s (such as when attaching the joists to the 4×4 sills), they are not strong enough (or long enough) to hold 2×4s together. The 6d or 8d nails also work fine for nailing in brace boards, floorboards, and sheathing.

Building Wall 2

The second wall, which will be 66 inches tall, will be shorter than the 6-foot width of the floor because it will be set between the two longer walls. These wall frames are each 3½ inches thick (the width of a 2×4), so they add up to a total thickness of 7 inches. Wall 2, then, will be 72"–7", or 65 inches wide.

FRAME WALL 2

Step 1. Set out two 2×4s that are at least 6 feet long on your sawhorses. These will be the top and bottom plates. Measure and cut them to 65 inches.

Step 2. Put the plates together, on edge, on your sawhorses. With your square and tape measure, measure from one end (it doesn't matter which end), and draw centerline marks at 24 and 48 inches on both plates. Set them aside.

Step 3. Set out four more 6-foot-long 2×4s on your sawhorses. These will be the studs. Measure and cut them to 63 inches. (If this wall is 66 inches tall, and the top and bottom plates together are 3 inches thick, then the studs will be 66"–3", or 63 inches long.)

Step 4. Lay the four studs on the ground with the plates so the center of each stud meets a centerline on the plates. Nail the plates to the studs as you did with the tall wall.

INSTALL WALL 2

Step 1. Get someone to help lift up and brace this wall next to the tall wall.

Step 2. Nail the bottom plate into the floor and the end stud into the end stud of wall 1. Use a level to get the wall plumb, having your helper gently nudge the wall as needed.

Step 3. Nail in another brace to hold it in place.

Two Walls Up

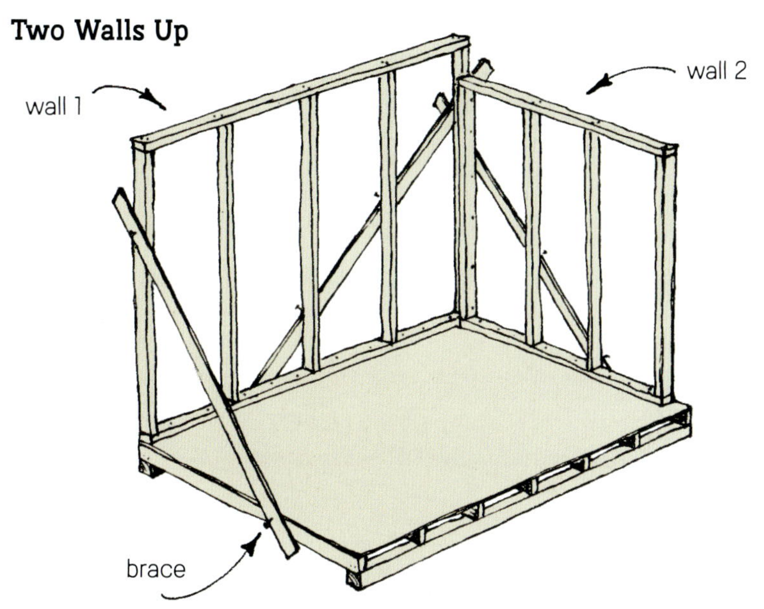

Building Wall 3 (with a Door and Window)

The third wall will be the same size as wall 2, and it will have the door and a window. You'll need to know the sizes of these openings before building the wall. In this example, you will build openings to fit a door 24 inches wide and 59 inches tall. If you have different door and window sizes, simply substitute the correct measurements.

Centerlines, Edge Lines, and X's

House builders use their own language to mark the position of studs and other pieces of wood in wall framing. The simplest mark is the centerline—a line to mark the center of the stud where it meets the plate, which we used in walls 1 and 2. Another mark is an edge line with an X. For wall 3, we are using edge lines and X's to tell us more accurately where the pieces go to fit the door and the window.

CUT AND MARK THE PLATES

Step 1. On your sawhorses, cut two 2×4s to 65 inches long, as you did for wall 2. These will be the top and bottom plates.

Step 2. Set the plates together, on edge, on your sawhorses. From the left end, measure to 25¾ inches, draw a line across both plates, and draw an X to the *right* of the line on both plates.

Step 3. From the right end, measure to 21¾ inches, draw a line across both plates, and draw an X to the *left* of that line on both plates.

Step 4. Draw an X at each end of the plates. The X's are where the studs go.

Step 5. Write a "T" for "top plate" on the upper 2×4, and a "B" for "bottom plate" on the lower 2×4 (see the drawing below). Once the boards are marked, there should be a 24¼-inch-wide space for the door at the left, and a 20¼-inch-wide space for the window at the right, as shown in the drawing below.

Step 6. Set the plates aside.

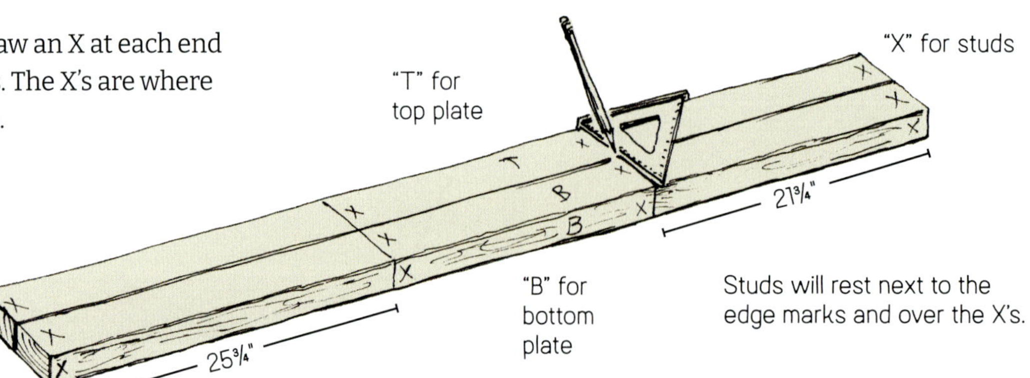

INSTALL THE STUDS

Step 1. On your sawhorses, cut four 2×4 studs 63 inches long, as you did for wall 2.

Step 2. Lay out the studs on the ground with the two wall plates, so that the studs all meet the plates at the X's. Nail the plates to the studs, stepping or kneeling on each stud to hold it in place.

Step 3. For the blocks that frame the door and window openings, measure 6 inches down all four studs from the top of the wall plate, then draw an edge line and an X above the edge line. These four X's show where the door and window header blocks will go.

Step 4. Along the two right-hand studs, measure another 25¼ inches from the first line you drew, then draw another edge line and an X below that line. These two X's show where the window rough-sill block will go.

INSTALL THE BLOCKING

Step 1. Cut two 2×4s to 20¼ inches long. These will frame the top and bottom of your window.

Step 2. Set the blocks in place beside the studs at the X's you just drew. You should end up with a window opening that is 25¼ inches high by 20¼ inches wide.

Step 3. If your window is 25 × 20 inches, drop it in gently to see if it will fit. It does? Great! Now take it out. If it doesn't fit, push on a corner of the wall frame to make it more square, then try again. If it still doesn't fit, you may have to adjust the blocks or move a stud.

Step 4. Cut one more 2×4 to 24¼ inches long. This will frame the top of your doorway. If you have a door already, see if it will fit and adjust as needed. Otherwise, you can build a door to fit later.

Step 5. Nail these blocks to the studs by using two 12d or 16d nails through each stud into the blocks.

Laying Out Wall 3

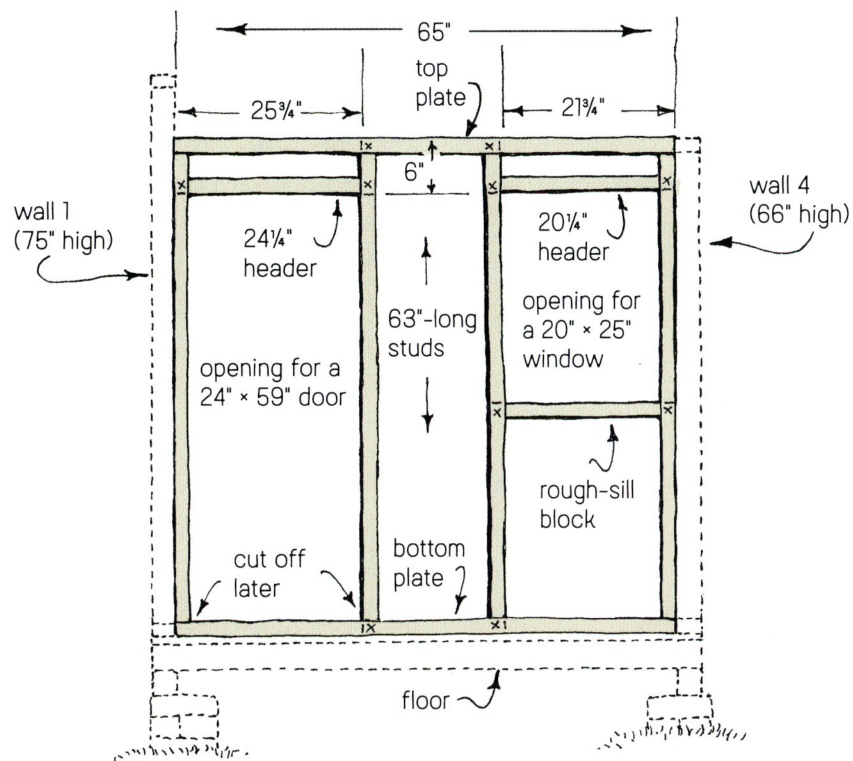

INSTALL WALL 3

Step 1. With a helper, set this wall in place and brace it as you did the others.

Step 2. Nail it down, first into the floor and then into the adjoining wall corners. Do *not* nail it to the floor in the doorway, where you'll cut out the bottom plate. (In fact, once wall 3 is up, you can use your handsaw to cut the bottom plate inside the doorway at any time.)

Step 3. Plumb the wall with your level while your helper gently nudges the frame as needed, then brace it in place.

Three Walls Up

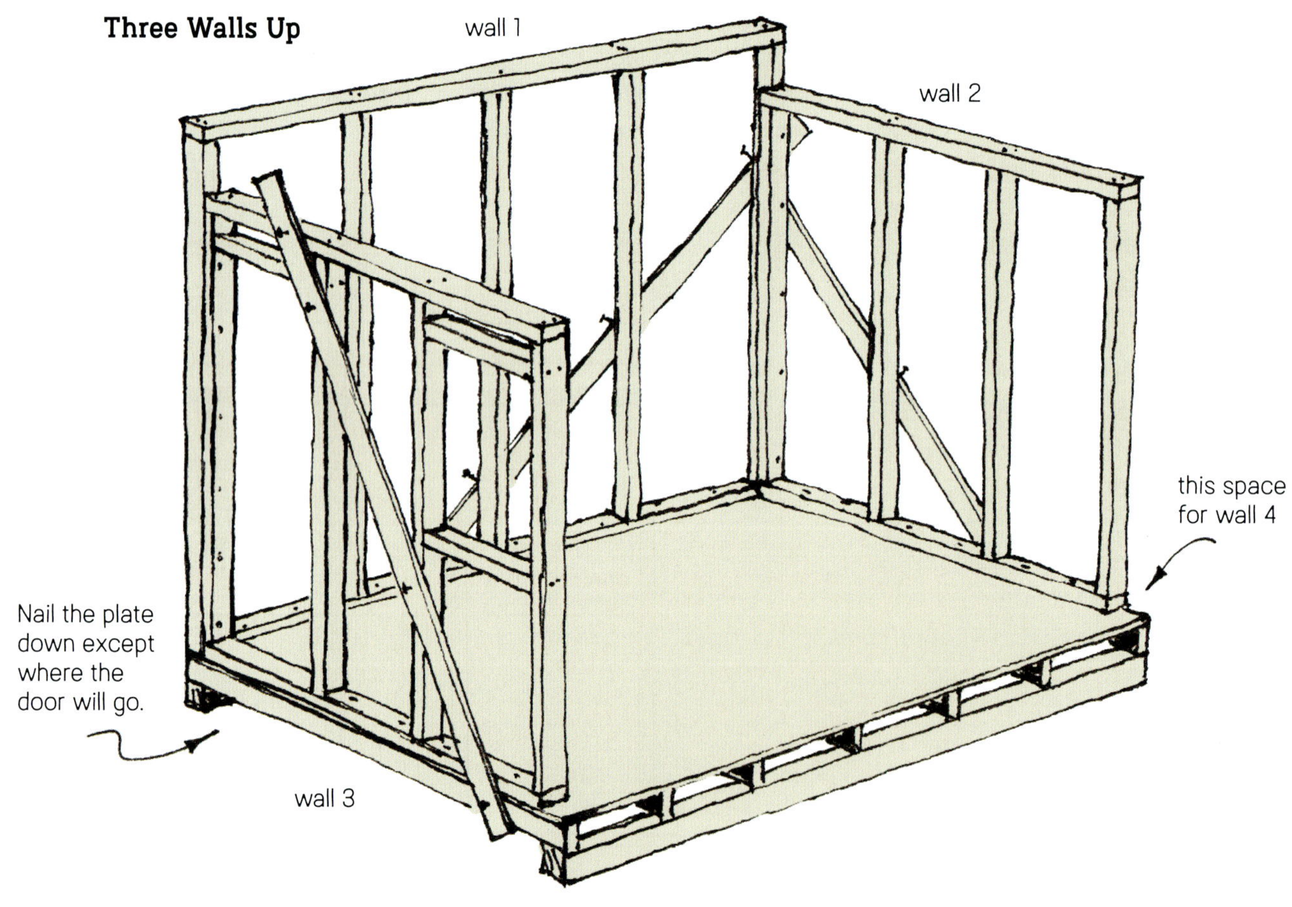

Building Wall 4

Wall 4 is also 66 inches tall and 8 feet (96") long. This wall will have one window, which I will show you how to frame in three different sizes.

FRAME WALL 4

Step 1. Lay out two 2×4s on your sawhorses for your plates and cut them to 96 inches long.

Step 2. Set them on edge side by side, then measure out and mark every 24 inches, as before. Draw the centerlines on both plates, then set the plates aside.

Step 3. Lay out five more 2×4s for studs, then measure and cut these to 63 inches long.

Step 4. Nail the plates and studs of this wall together as you did the others, and leave it on the ground.

FRAME THE WINDOW

On this wall, you will frame the window in a slightly different way than on wall 3.

Step 1. Choose the two studs that will surround your window. On each of these two studs, measure and draw an edge line 6 inches down from the top of the wall, then draw an X above the line.

Step 2. Measure another 25¼ inches from the first edge line, draw another edge line, then an X below that line. These two X's show where the window rough-sill block will go (see the drawing on page 58).

Step 3. Measure the space between the two studs, which should be about 22½ inches. Cut two 2×4 blocks that long, set them on the X's, and nail them into the frame. These blocks are your window header and rough sill.

Step 4. Measure to 20¼ inches on both the header and rough sill, draw an edge line, then draw an X beyond the edge line.

Step 5. Measure and cut another 2×4 block, called a trimmer, to 25¼ inches long.

Step 6. Nail the trimmer block in place on the X's (see the drawing on page 58).

Window Safety

For safety, use only windows in wood or metal frames, instead of just unframed sheets of glass. Also, the bottom of any window should be far enough above the floor (at least 12 inches) so that someone doesn't accidentally kick it or crash into it.

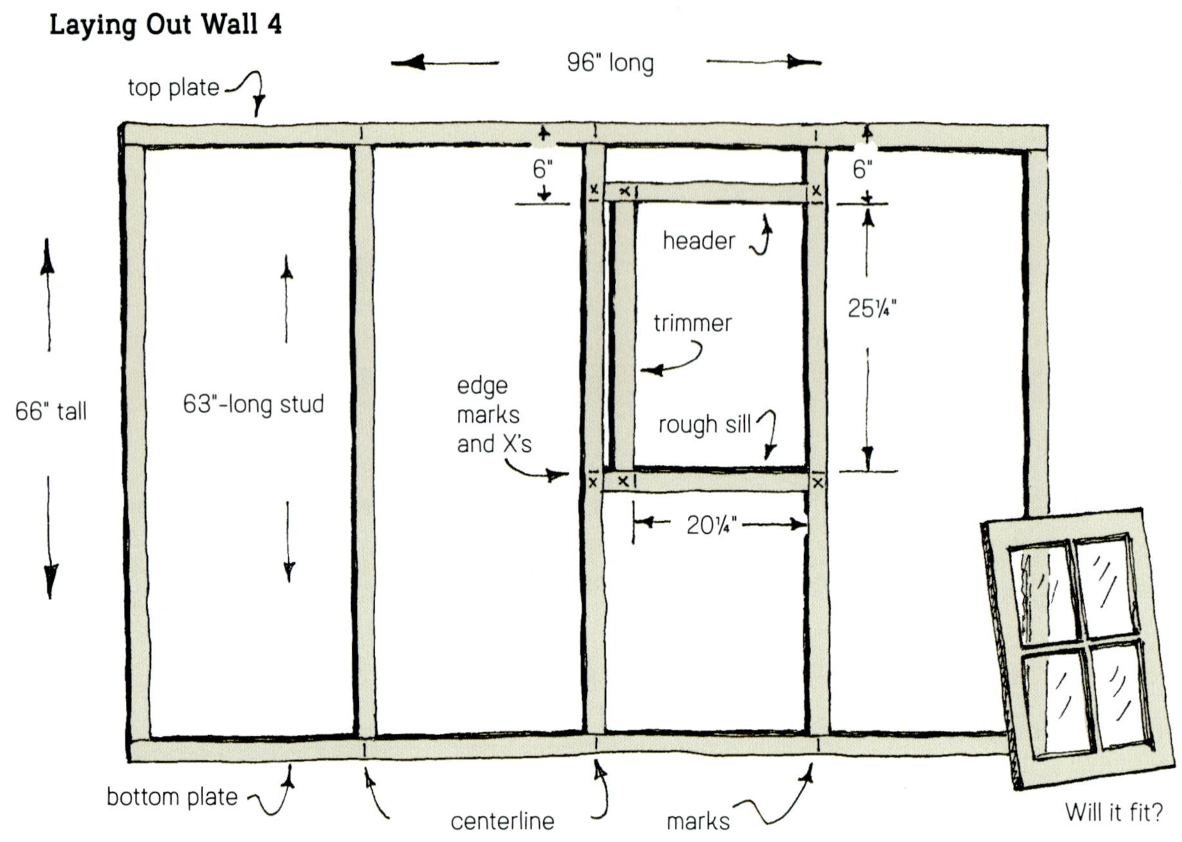

Bonus: How to Frame Bigger Windows

If you have to cut through a stud to accommodate a big window, frame the wall as shown on page 59 at top right, with a full stud to each side of the window opening. The short stud pieces above the window header and below the window rough sill are called cripples.

If you have to cut through two studs to accommodate an even bigger window, frame the wall as illustrated on page 59 at bottom right. The header should be a 4×4 or two 2×4s nailed together, and it should be long enough to span the window opening plus the thickness of the two trimmer pieces that hold it up.

Wall 4 with a Big Window

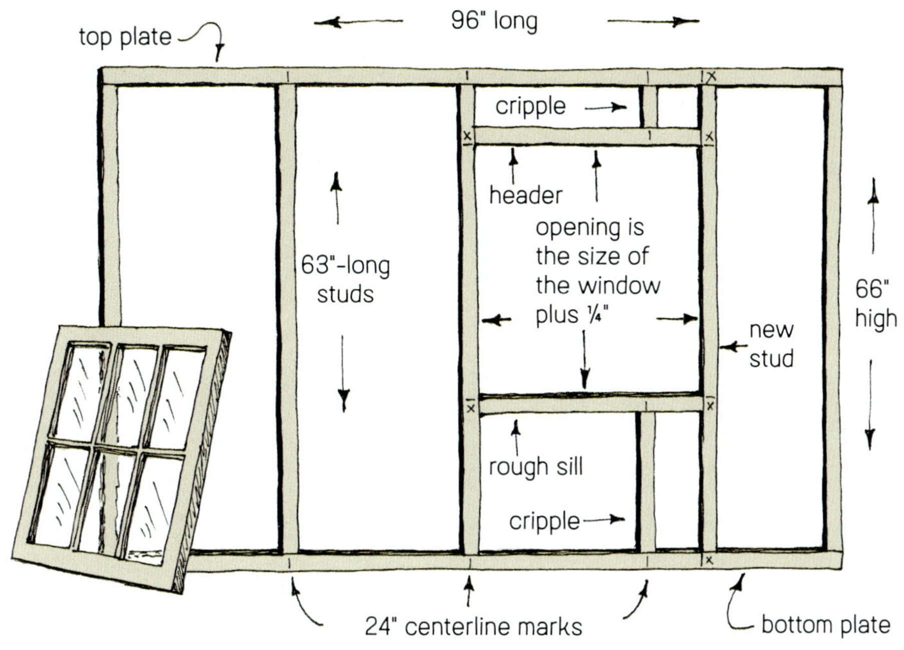

Wall 4 with a Bigger Window

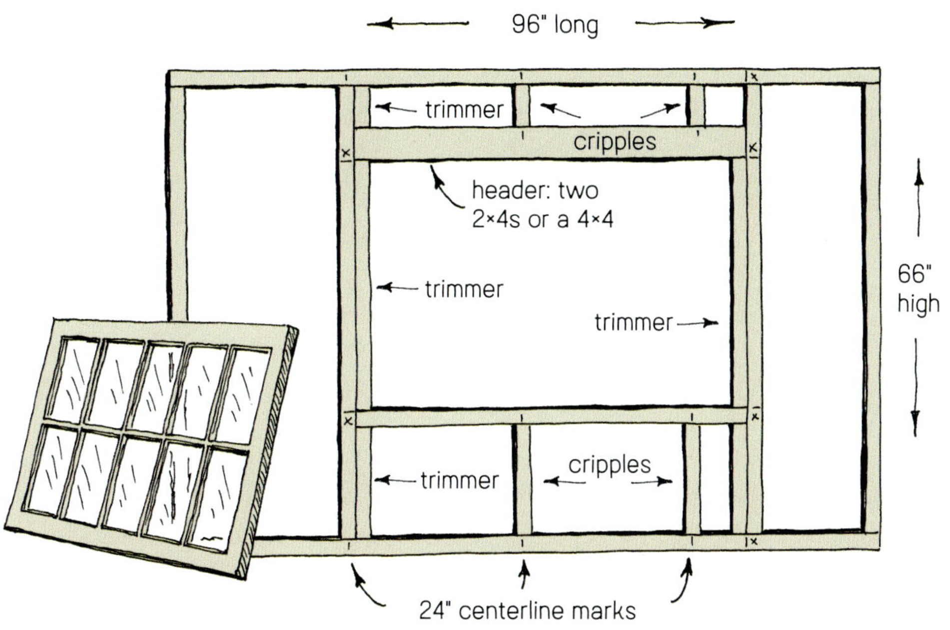

INSTALL WALL 4

Step 1. With your helper, set this wall in place and brace it as you did the others.

Step 2. Nail it down, first into the floor and then into the adjoining wall corners.

Step 3. Plumb it with your level while your helper helps gently nudge the frame, then nail in another brace board.

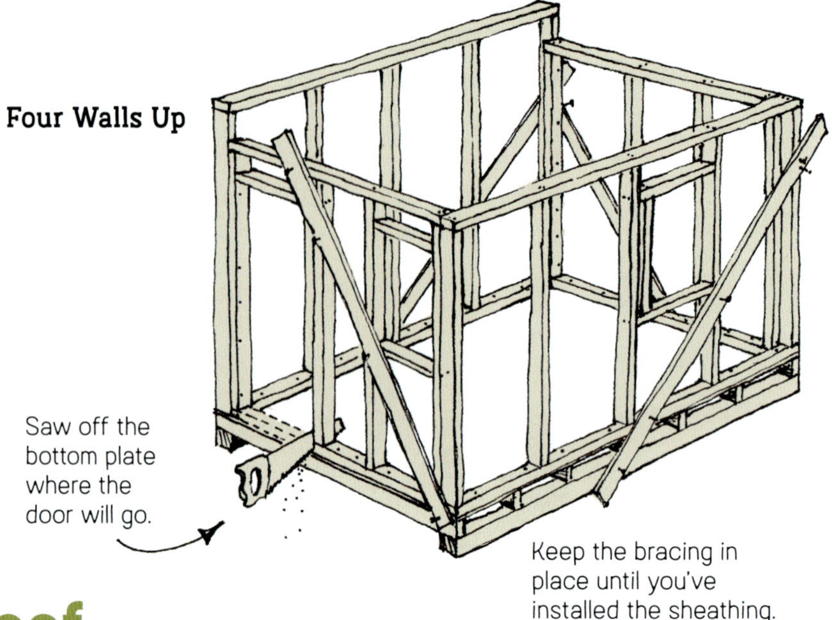

Framing the Roof

Now we'll need to put up two beams—a 2×6 and a 2×4, both 8 feet long—to hold up the roof boards.

CUT AND INSTALL THE ROOF BEAMS

Step 1. Grab a stepladder and climb up to look at the top of one of your 6-foot-wide short walls. Hook your measuring tape to the outer edge of the tall wall, measure to 24 inches along the top of the 6-foot wall, and draw a centerline. Then pull the tape farther along and draw another centerline at 42 inches.

Step 2. Go around to the other short wall and draw similar centerlines the same distances from the tall-wall edge. These will mark the approximate places for the beams.

Step 3. Set the 2×6 and 2×4 on your sawhorses. Measure them; if they're not 8 feet long, cut them to length. These will be your roof beams.

Step 4. With an assistant, haul the roof beams up to the roof and stand them on edge on top of your short walls, aligned over the X's. The 2×6 will be the upper roof beam, positioned closer to wall 1. The 2×4 will be the lower roof beam, positioned closer to wall 3. Don't nail them in just yet.

Step 5. Lay a straight test board, or your long level, across the beams and the wall tops. Adjust your two beams by nudging them until the board rests on both of them. If you move the beams off the centerlines you drew, that's okay.

Step 6. When you have the beams positioned so that the board just touches the wall 1 top plate, the two roof beams, and the wall 3 top plate, toenail the beams into the top plates of walls 2 and 4, as shown in the drawing on page 61.

Framing the Roof

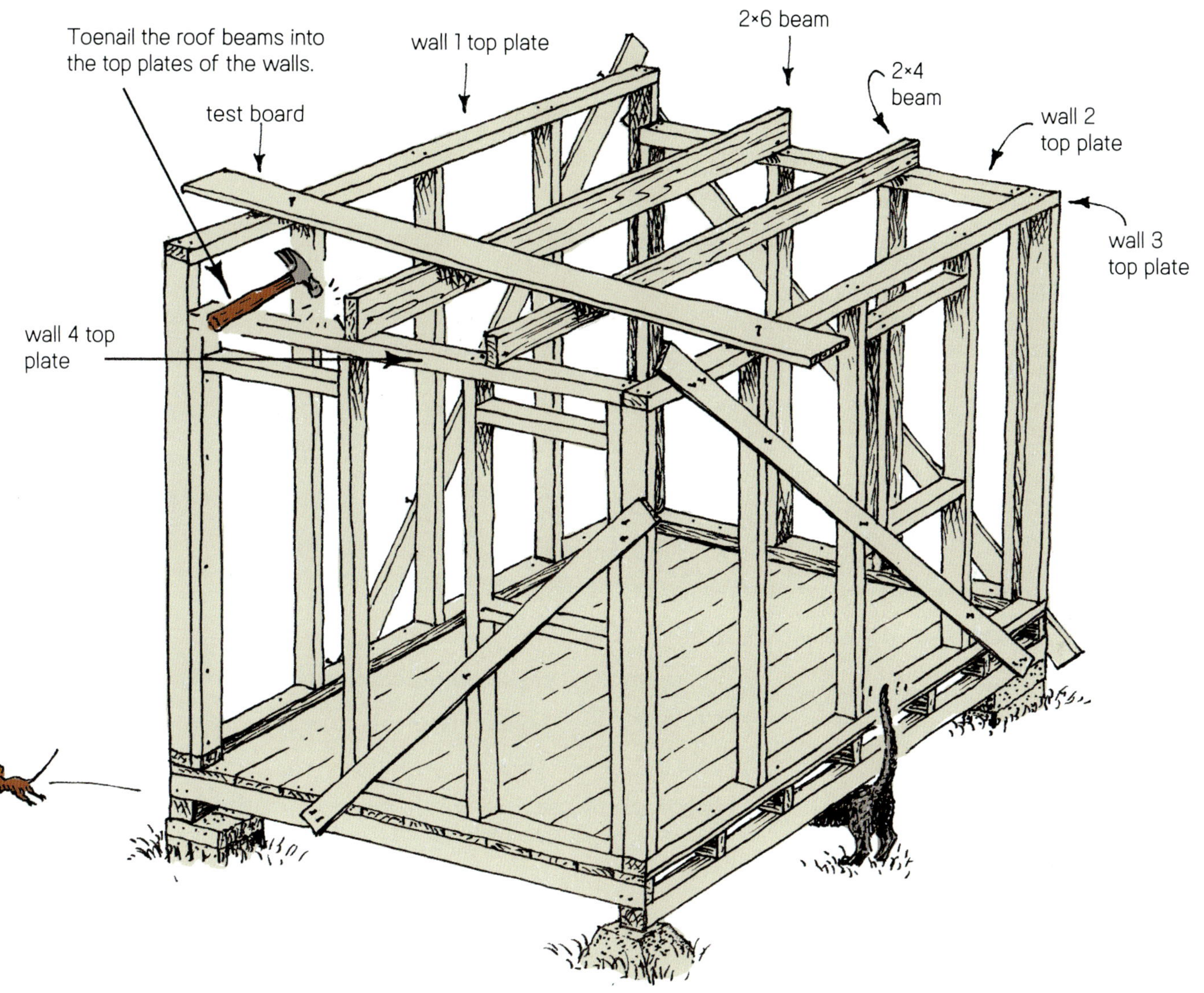

·CHAPTER 5·

Finishing

Now it's time to enclose your structure with sheathing, which can be boards, plywood, or OSB panels. Then you'll install the door, windows, and trim boards. Again, you'll need your essential tools, along with a utility knife, a stepladder, and the materials listed below.

FINISHING MATERIALS

PART	QUANTITY	DESCRIPTION
WALLS	65	1×6 car siding or pine boards, 8 feet long; *or*
	5	4 × 8-foot sheets 3/8-inch-, 7/16-inch-, or 1/2-inch-thick CDX plywood; *or*
	5	4 × 8-foot sheets 3/8-inch- or 7/16-inch-thick OSB
ROOF BOARDS	20	1×6 car siding or pine boards, 8 feet long; *or*
	14	1×8 car siding or pine boards, 8 feet long; or
	2	4 × 8-foot sheets 3/4-inch-thick plywood or OSB
ROOF COVERING	1	roll mineral-coated roofing (better) or 30-pound felt underlayment (okay)
TRIM BOARDS	2	1×3s or 1×4s, 10 feet long, for roof edging
TRIM BOARDS	14	1×3s or 1×4s, 8 feet long, for corners, doors, and windows
WINDOWS	2	20 × 25-inch barn sashes, or other size, depending on what you find
WINDOW HINGES	2–4	small utility or old cabinet hinges, with screws
DOOR	1	recycled or salvaged closet or other small door, or build one yourself (see how to build a door on page 66)
DOOR HINGES	2	6- or 8-inch-long T hinges, or 3 × 3-inch butt hinges, with screws
NAILS	2 pounds of each	8d coated sinkers; 6d coated sinkers
NAILS	1 pound of each	6d galvanized box nails, for the trim boards; 6d duplex nails, for the brace boards; 3/4-inch- or 1-inch galvanized roofing nails
SCREWS	1 pound	1 1/4-inch Phillips-head construction screws (optional; useful if you're building your own door)

Covering the Walls

The sheathing is the exterior covering of the structure, and it also strengthens and braces the framing. So take off the brace boards only when you are about to cover the wall they are bracing. Here are some tips:

- Nail on sheathing boards across the studs starting at the bottom of each wall. By "bottom," I mean that the sheathing should cover the 2×4 joists but not the 4×4 sill.
- Always join boards or panels over a stud so the ends or edges are nailed to something solid.
- Use 8d nails for boards, driving in two or three nails at the ends of each board and two nails on every inside stud. Use 6d nails for ⅜-inch plywood or OSB sheathing, and 8d nails for thicker sheets. Nail them every 6 inches on the edges and every 12 inches on the inside studs.
- Install sheathing right up to the edges of window and door openings, notching it as necessary so it sits flush with the edges of the openings.
- When you get to the top of a wall, mark the last board with a pencil line, saw off the extra wood, then nail up the board.
- To mark the top of a board on the slanted wall, put a roof board against it, and mark the wall board edge minus the thickness of the roof board (see the drawing below, left).

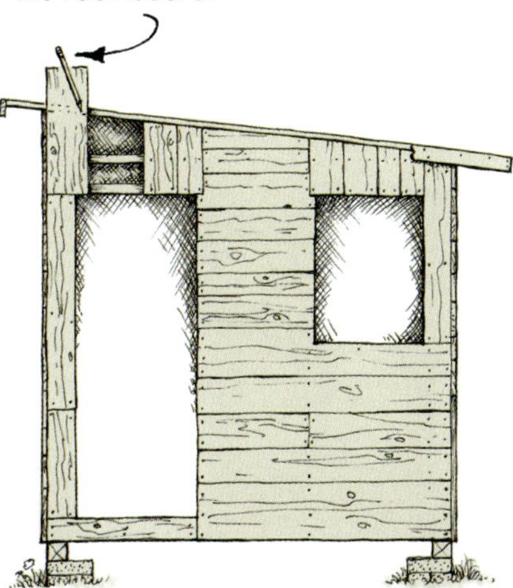

Sheathing a Wall with a Slanted Top

Sheathing and Siding in One

Car siding and board-and-batten (see page 95) are two materials that can serve as both sheathing and finish siding. If you have to buy new sheathing, these might be the best options, since they aren't very expensive and can double as your finished outside siding. Car siding works equally well for the floor and the roof.

Ready-to-paint plywood or engineered exterior panels can also serve as both sheathing and finish siding. If you want to use these for finish siding, carefully measure it so the edges of the panels are centered over studs and the grooves are always outside and vertical for a good look.

See Chapter 7 for more on finish siding and how to install it.

Building the Roof

A flat roof will leak, even with tar paper on it, so your roof should be slanted and extend beyond the walls to help keep them dry. For the roof boards, I recommend using pine boards or 1×6 or 1×8 car siding. The boards should be about 7 feet 6 inches (90") long to allow for an overhang at both ends. For added strength around the edges of your roof, add 1×3 or 1×4 trim boards.

INSTALL THE ROOF BOARDS AND TRIM

Step 1. Set your roof boards on your sawhorses and cut them all to 7 feet 6 inches long.

Step 2. On each roof board, measure and mark a line 8 inches in from one end, to mark the spot where the overhang will start.

Step 3. Place one of the boards across the roof, on one of the outside edges, slanting down from the tall wall to the opposite wall. Align the overhang mark at the outside edge of the short or tall wall. Align the long edge of the board so that it covers the top edge of the sheathing on the side wall. (If there is no sheathing yet, tack on a ¾-inch-thick piece of scrap to the wall framing to use as a guide.)

Step 4. Nail the roof board into the top wall plates and the roof beams with 8d nails.

Step 5. Repeat the process with another roof board at the other side of the roof.

Step 6. Nail on all the remaining roof boards, being sure to align with the overhang marks. When it comes to installing the last board, you may find yourself with a gap that the board doesn't quite fit into. In this case you'll have to rip the board (cut it the long way) or find a narrow board to fit.

Step 7. Using 6d galvanized box nails, nail on trim boards all around the edges of the roof. Use the 10-foot-long 1×3 or 1×4 furring strips on the long edges of the roof, and the 8-foot-long furring strips on the shorter walls, cutting them to size as necessary. (If you haven't yet installed the wall sheathing, though, hold off on installing the trim on the short walls until after you've nailed on the sheathing.)

Roofing with Car Siding

If you are using car siding, mark the boards for the overhang as described in Step 2. Trim the tongue off the outer edge of one board with your handsaw, then nail the board on at the edge of the roof as described. Working your way across the roof, nail on the remaining boards with the grooved side facing down, joining them one at a time. As you approach the other edge of the roof, rip (cut) the last board lengthwise to make it fit, and then nail it down. Add the trim boards as described. (For more on installing car siding, see page 93.)

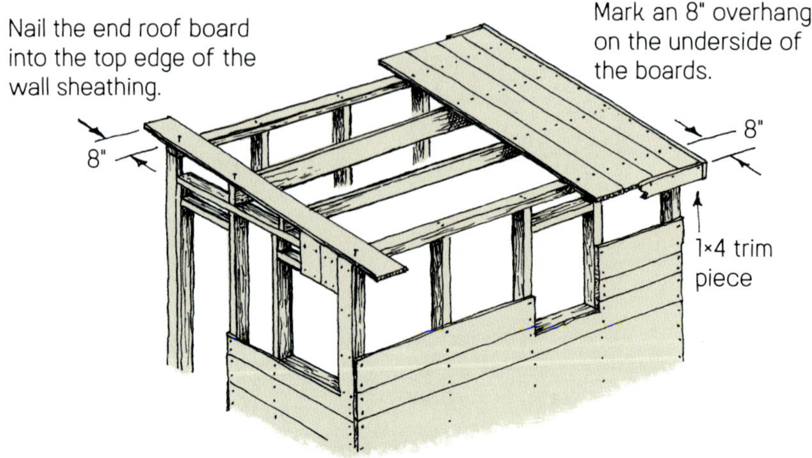

COVER WITH ROLL ROOFING

Cover your roof with 30-pound felt (tar paper) or, for a more permanent roof, mineral-coated roll roofing. One roll should be enough. (You can also use 15-pound felt with asphalt shingles, which I talk about in Chapter 11.)

Step 1. On the ground, roll out a length of roll roofing and measure a piece that will be 2 inches longer than the length of your roof. Measure the roof to double-check.

Step 2. Cut off the piece with a utility knife, using your square to set a board to cut against so you get a straight edge. Cut on the back side if you are cutting mineral-coated roofing.

Step 3. Roll up the piece and carry it up to your roof. With a helper holding your ladder, set the piece of roofing along the lower edge, overhanging the trim board by 1 inch. This allows for a "drip edge."

Step 4. Nail the roofing down with ¾-inch- or 1-inch-long roofing nails. Use one nail every 4 inches along the bottom and side edges, and one nail every 12 inches or so along the top edge.

Step 5. Cut and nail in a second piece of roofing, overlapping the first piece by at least 4 inches.

Step 6. Repeat with a third piece, allowing 1 inch to hang over the top edge.

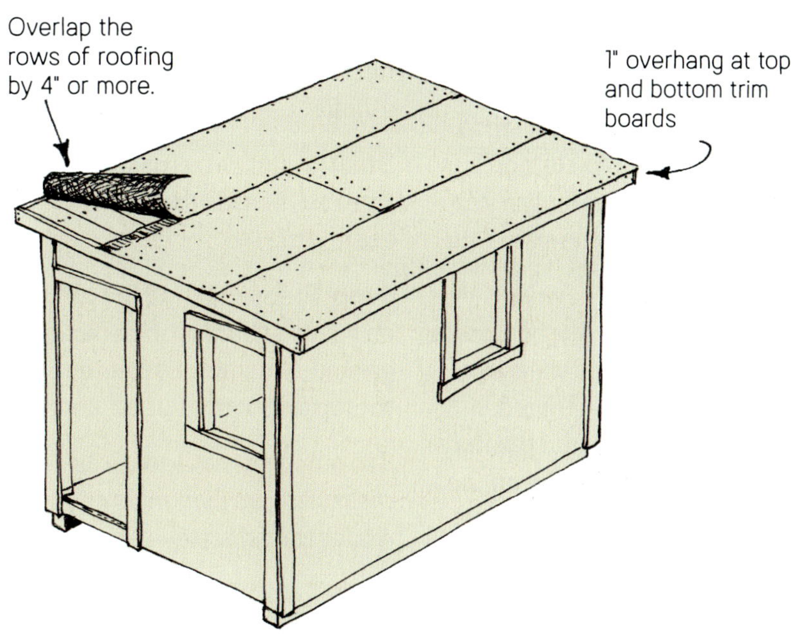

Putting in the Door

Maybe you found a cool door at a recycled building-supply store or in an alley. An old-fashioned tall cabinet door might work, or even a piece of plywood. Lumberyards often sell plywood in small sizes, typically called handy panels. Ask for a 2 × 4-foot panel that is ¾ inch thick, which will make a good small door if you have children. If you can't find a door that fits, you can always build your own door.

Whether you build your own or find a door, you'll need two hinges. Old door hinges or gate hinges will work fine. If you buy hinges, look for 6-inch-long T hinges, which are shaped like a T and meant for gates and shed doors (see the drawing on page 37). You'll also need ¾-inch- to 1¼-inch-long screws to hold the hinges tight (nails will loosen up after a while).

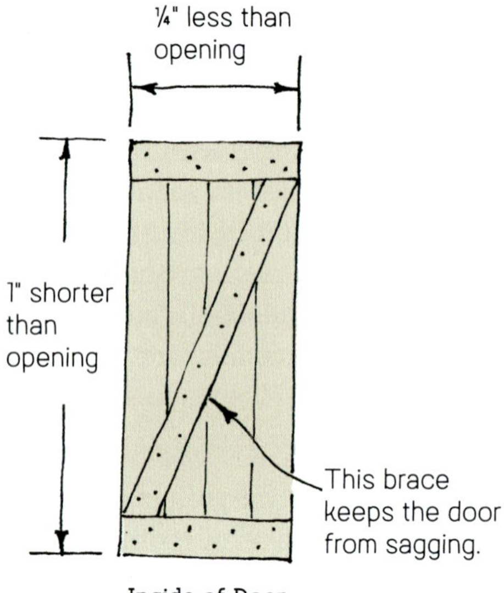

BUILD YOUR OWN DOOR

You can build your own door using 1×6 boards or car siding. Here's how.

Step 1. Measure your door opening, then subtract ¼ inch from the width (side to side) and 1 inch from the length (top to bottom); this is the size of the door you need.

Step 2. Lay out a few boards on your sawhorses and measure the exact width you'll need. If you're lucky, you'll find the right width of boards to make up the door. Otherwise, first cut the boards to length, and then mark one of the boards to be ripped, or cut lengthwise, to make up the width you need. Use a straight board to draw this long cut line.

Step 3. Along the cut line, rip, or saw off, the extra part of the board. Be patient and get some help if you need it; it might take a while.

Step 4. Cut two crosspieces from 1×6 boards the width of the door.

Step 5. Assemble all the pieces on your sawhorses or on the ground, with the crosspieces over the top and bottom of the door.

Step 6. Square the door by measuring its diagonals as you did for the floor (see page 44).

Step 7. For the brace, lay another board diagonally over the crosspieces, from one edge of the door to the other.

Step 8. Mark cut lines at the points where the brace board crosses the crosspieces, then saw it to size.

Step 9. Nail or screw all the pieces together, being careful to keep the door square. This is where the 1¼-inch construction screws can be used, for which you'll need a drill. Otherwise, drive 6d galvanized box nails through the door and bend over the pointed ends that stick out past the wood.

Note: You can also make a door from plywood. Cut it to the size specified in step 1. If it is less than ¾-inch-thick, add cross-pieces at the top and bottom to hold the hinge screws.

HANG THE DOOR

Step 1. Lay the door flat, then set the top hinge 4 to 6 inches down from the top of the door and the bottom hinge 6 inches up from the bottom of the door. Use a pencil to mark the screw holes for each hinge.

Step 2. If you have a drill, use it to drill a pilot hole (a guide hole for a screw) at each mark, using a drill bit that's a little narrower than the screws. If you don't have a drill, use a nail (again, just a little narrower than the screws), pounding it partway into each screw-hole mark and then pulling it out.

Step 3. Holding the hinges in place, screw in the screws. Be patient and keep at it!

Step 4. With help from an assistant, set the door in place and wedge it with some shingles or shims, especially under the door, so it won't stick when you close it. A door should have ⅛ to ¼ inch of wiggle room on the top and sides and a ½- to ¾-inch gap on the bottom.

Step 5. Mark the location of the hinge screw holes on the adjacent wall, then make pilot holes, as you did on the door.

Step 6. Screw the hinges into the wall, and voilà! You have a door.

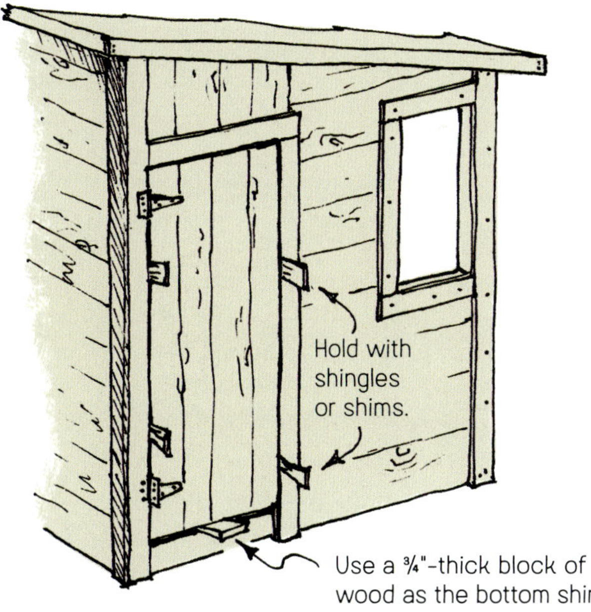

INSTALL THE KNOB, LATCH, AND DOORSTOP

Step 1. Once the door is hung on its hinges, put a handle or a knob on the outside of the door so you can open it, and another handle on the inside so you can pull it closed.

Step 2. So you can keep the door closed, put a latch on the outside of the door. The easiest way is to nail or screw on a piece of wood that you twist to hold the door shut. You can also add a hook-and-eye latch to close it from the inside.

Step 3. Have someone (that you trust!) latch the door on the outside while you are inside. While the door is latched, nail on a piece of trim board, called a doorstop, onto the doorframe, right up against the inside of the door. This will keep the door from opening too far inward and ripping off the hinges.

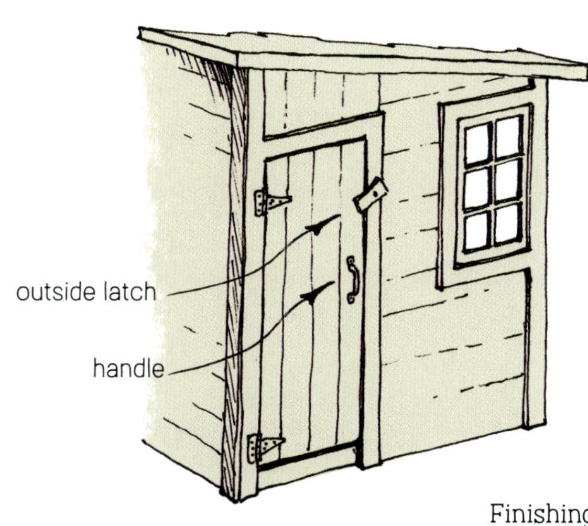

Putting on the Trim Boards and Windows

The trim boards not only make your backyard dwelling look good but may also help hold in your windows. A "fixed" window is one that doesn't open, so it doesn't require hinges. A hinged, openable window is nice for ventilation.

TRIM THE CORNERS

Step 1. At each corner of the structure, measure from the bottom of the sheathing to the bottom edge of the roof to find the length of trim you'll need. Cut trim boards to length from 1×3 or 1×4 boards. On the slanted walls, cut one end of the trim board at a slight angle to fit snugly against the roof trim.

Step 2. Nail on all the trim boards with 6d galvanized box nails.

INSTALL A FIXED WINDOW

Step 1. Measure and cut the trim boards so that they will overlap the window opening by ½ inch on every side. Nail in place with 6d galvanized box nails.

Step 2. Set the window in the frame from the inside. Nail more trim boards on the inside to hold the window in.

Note: Be careful to fasten the inside trim boards only into the structure's framing and not into or through the window frame itself.

INSTALL AN INWARD-OPENING WINDOW

You can install an inward-opening window directly to the wall frame stud, or to the inside trim as shown on page 69.

Step 1. Screw the hinges to the window frame.

Step 2. Set the window in its opening using thin wooden shims, as you did for the door (see page 67).

Step 3. Mark the screw holes for the hinges on the framing and then take the window out.

Putting in a Fixed Window

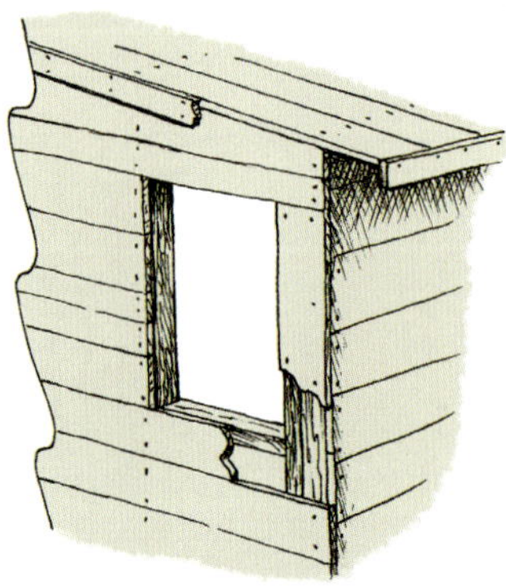

Sheathing is flush with the opening.

Trim boards overlap the outside edges of the window opening by ½".

Inside trim boards hold the window in place.

Step 4. Carefully make the pilot holes for the screws in the framing, then screw the hinges into it.

Step 5. Add a small trim board to the inside of the window opening (like the doorstop) to keep the hinges from ripping off.

Step 6. Add a cabinet knob or handle to open the window, and a latch to lock it.

If the Window Will Open Inward . . .

Hinge to the edge of the stud . . .

to the side of the stud . . .

or to the inside trim.

INSTALL AN OUTWARD-OPENING WINDOW

Step 1. Install the trim boards so they are flush with (rather than overlapping) the window frame. For now, leave off the trim board on the edge where the hinges will go.

Step 2. Set the window in place using some shims so it is even with the edges of the trim boards.

Step 3. Position the hinges where you want them. You can put hinges on the side of the window, but it is better to hinge an outside-opening window at the top so the wind doesn't blow it closed. Mark the screw holes with a pencil, then take out the window.

Step 4. Make all the pilot holes, then screw the hinges into the window frame.

Step 5. Put the window back in and screw the hinges into the sheathing, as shown.

Step 6. Nail on the remaining trim board over the hinges.

Step 7. Put a narrow trim board on the inside of the window opening (like the doorstop) to stop the window from opening too far in and ripping off the hinges.

Step 8. Add a prop stick to hold the window open and a hook-and-eye latch on the inside if you want to lock the window.

If the Window Will Open Outward . . .

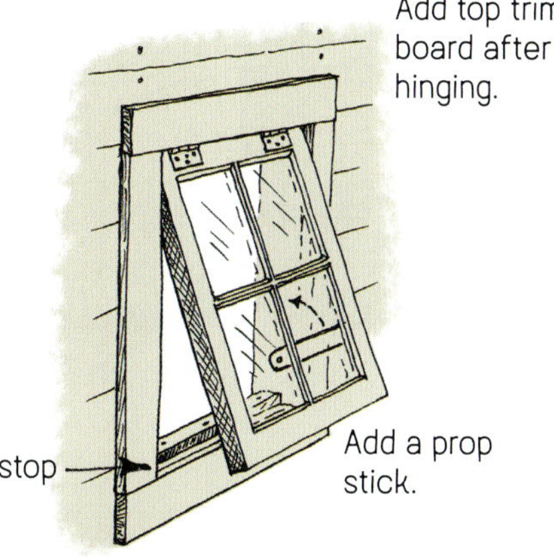

·CHAPTER 6·

Building It Bigger and Better

You thought you were done, right? Well, maybe you are . . . but say you have a lot of visitors now, since backyard gathering spaces seem to be powerful magnets, and you need more room. Again, there are many possibilities! This chapter describes how to build inside walls and add more rooms.

Building an Inside Wall

A small room or alcove can be useful in a backyard dwelling. It's a place where you can read or work or gather with friends and family. You can create your own private nook with an inside wall, also called a partition.

Sometimes building an inside wall might seem like a great idea, but once it's up, it might make the space feel crowded, dark, or awkward. If you are not sure you want a partition, try hanging an old blanket where your wall would be to see how it feels.

BUILD A SIMPLE PARTITION

You can build a simple partition by nailing 2×3 or 2×4 studs to the roof beams and to the floor, then nailing boards across the studs. Here's how.

Step 1. Decide how wide you want your new room to be.

Step 2. Measure along each roof beam from an existing wall, and make a mark on the beam. Measure the same distance along the floor and make another mark.

Step 3. Measure from the ceiling to the floor to find the length of the studs.

Step 4. Stand each stud on the floor, nail it to the roof beam by your mark, and use your level to make sure each stud is plumb before nailing it to the floor.

Step 5. Toenail the studs into the floor.

Step 6. Nail a stud against the wall where the partition will meet it, or use one of the studs already in the wall, if you haven't paneled it yet.

Step 7. Nail boards to the studs, and you have your partition!

Alcove

BUILD A WALL-FRAME PARTITION

If you have already paneled your inside walls or you want real inside walls, you can build partition frames outside, then bring them in and nail them in place.

Step 1. Decide on how big your inside room will be.

Step 2. Measure and mark the ceiling beams and the floor where the walls will go.

Step 3. Carefully measure the distance between the floor and the ceiling beams so you know how tall your partition frames should be. Also check to make sure the frames will fit through the door.

Step 4. Build each wall frame outside, then bring them inside. Set each to your marks and nail them to the ceiling beams, then use your level to make them all plumb.

Step 5. Nail each wall frame into the floor and into the walls.

Step 6. Add boards or paneling to one or both sides of the frames to close up your new room (see more about installing inside paneling on page 97).

Step 7. Hang a curtain over the doorway or build a door to create a cozy inner sanctum.

Room with Doorway

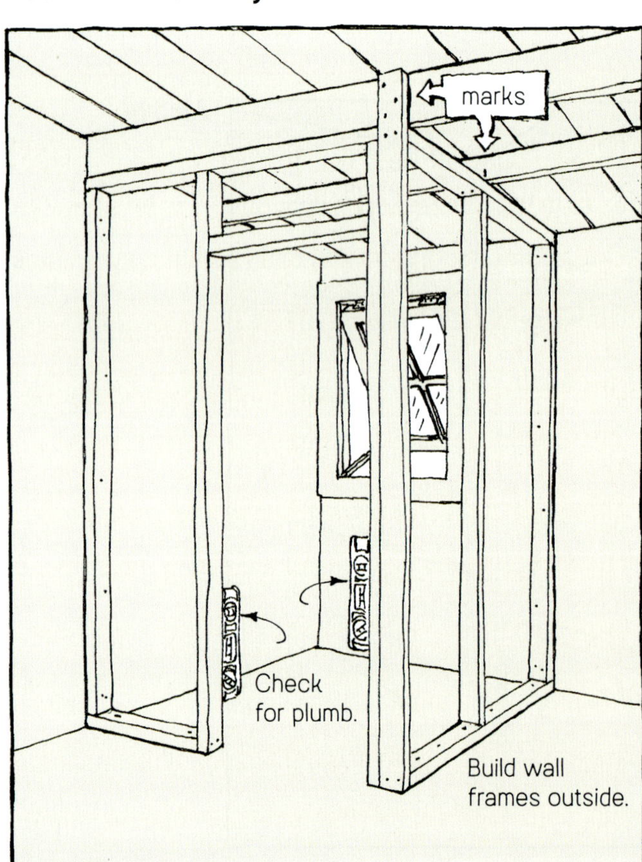

Partitioning Possibilities

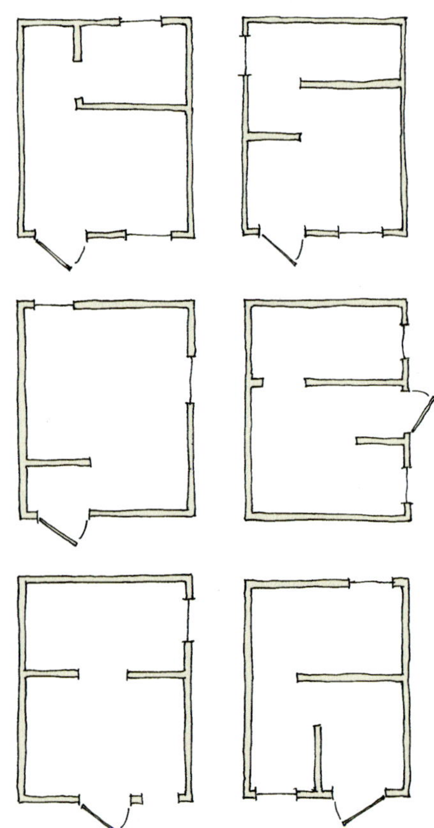

Adding a Room

If your dwelling becomes too small, you can always add another room. To determine how big your addition should be, think about what you'll want to use it for. Then think about how it will attach to your original design. Is there room for it? Draw floor plans and maybe a picture of how you think the new room might look. Trying out different ideas can save a lot of extra work in the long run.

I'll describe a room that's 5 × 6 feet in size, but of course you can design a different size or shape. To build it this size, you will just need your essential tools, a stepladder, and the materials listed on page 74.

Possibilities for Adding a Room

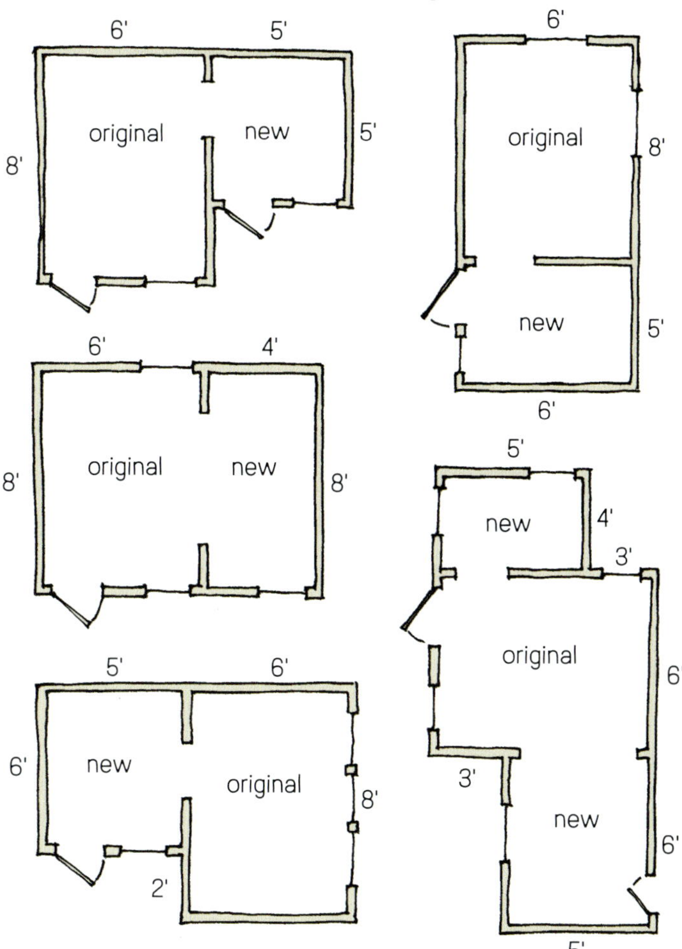

This is the plan we'll use as our example in the following instructions for adding on a room.

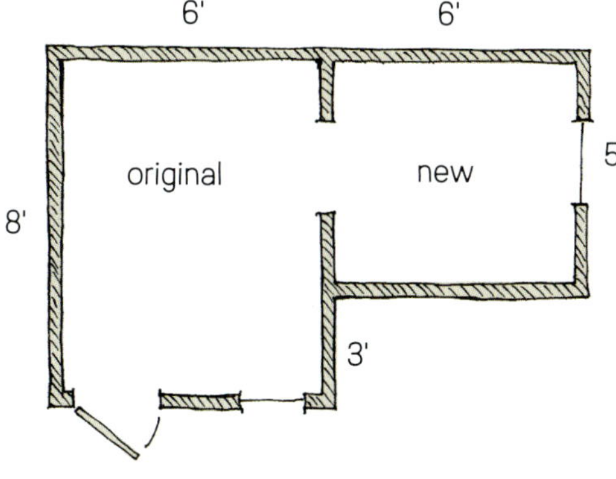

MATERIALS FOR ADDING A ROOM

PART	QUANTITY	DESCRIPTION
SILLS	1	4×4, 10 feet long (to be cut in half)
ROOF BEAMS	1	2×6, 10 feet long (to be cut in half)
JOISTS, STUDS, AND ROOF BEAMS	28	2×4s, 6 feet long
FLOORBOARDS, WALL SHEATHING, AND ROOF BOARDS	25	1×8 car siding boards, 12 feet long (pine boards, plywood, OSB, or other sheathing can be substituted)
ROOF COVERING	1	roll mineral-coated roofing (better) or 30-pound felt underlayment (okay)
BRACE AND TRIM BOARDS	8	1×3 or 1×4 furring strips, 8 feet long
WINDOW	1	any size, with hinges and screws (optional)
DOOR	1	any size, with hinges and screws (optional)
NAILS	2 pounds of each	12d or 16d coated sinkers; 6d or 8d coated sinkers
NAILS	1 pound of each	6d galvanized box nails; ¾-inch- or 1-inch-long roofing nails

BUILDING THE FOUNDATION

It's important that the new foundation be square to and level with the foundation of the existing structure. Here's how to lay it out.

LAY OUT THE FOUNDATION

Step 1. Clear the ground and lay out a foundation where you want to build your new room. Measure out 6 feet from your dwelling and 5 feet across, then drive stakes to mark your corners, as you did for your original design (see page 43).

Step 2. Check the stakes for square by measuring diagonally from corner to corner. At 6 × 5 feet, your new foundation will be square if your diagonal measurement is 93¾ inches in both directions.

Step 3. Lay cornerstones of concrete blocks or bricks at all four corners. Set them so they are level with those under your structure, so that your new floor will be at the same height. If the ground slopes, set your new cornerstones level to each other, and allow for a step up or down to your new room.

BUILDING THE FLOOR

Build the new floor with sills, joists, and floorboards as shown in Chapter 3 (see pages 42–49).

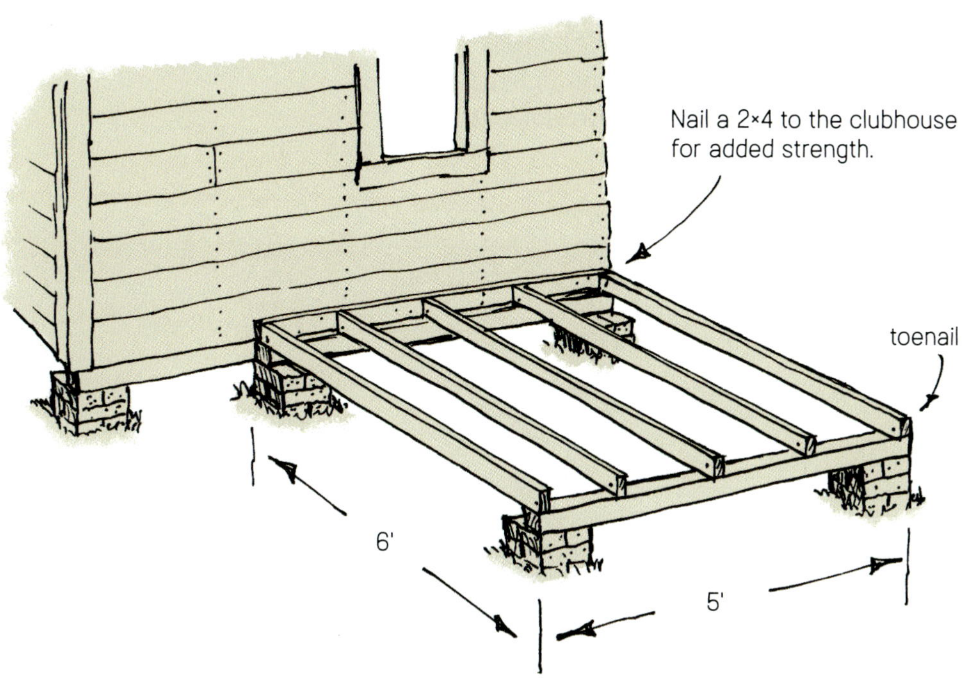

Reality Check

Remember to contact your local building-inspection department and your neighborhood owners' association before building an addition. Once you get the thumbs-up, play around with drawing what an addition might look like. Have fun with it!

To help you with planning, page 73 provides some possible floor plans and drawings showing how an addition could connect to the Classic Design. To minimize the work needed to build a doorway, you can add the room to that part of your dwelling where there is already a window or a door. You can then build a new door on your addition. Also, think about adding a window or two for more light and air. Otherwise, build the addition the same way you built your original structure.

MAKING THE INTERIOR DOORWAY

Now would be a good time to make a doorway from your original structure to your new room. First comes deconstruction.

CUT A NEW DOORWAY

Step 1. If there is a window in the wall, take it out and saw off all the sheathing and paneling down to the floor below the old window. Watching for nails, use a crowbar to pry out any blocking. If there is no window, take out one board, preferably at the top of your new doorway, and saw out a doorway along the side of two studs, all the way to the floor. You may have to remove a larger piece of plywood or OSB sheathing to begin sawing, then replace the scraps.

Step 2. Saw off the bottom plate that passes through the new doorway so you won't trip over it.

HOW TO FRAME A WIDE DOORWAY

If you want a wide opening through your original wall into your addition and you have to take out a stud or two, then you'll have to support the original wall with a beam called a header. This is the same kind of header you would put in over a big window (see page 59). Here's how to frame a wide opening to your new room.

Step 1. Remove the sheathing, studs, and bottom plate from the original wall. Leave the top plate in place.

Step 2. Cut two 2×4 trimmers to the height of your new opening and nail them to the remaining original studs.

Step 3. Cut a header long enough to rest on the trimmers and fit snugly between the original studs. A 4×4 header will work fine, or you can make a header by nailing together two 2×4s with some scraps of ½-inch-thick plywood or OSB in between to make it the thickness of a wall.

Step 4. If there is room, add short 2×4 pieces (called cripples) above the header to support the top plate.

Step 5. Replace any sheathing you removed on the original wall (see the drawing below).

A Superwide Opening

If the opening into the addition is so wide that the header will be more than 5 feet long, use a solid 4×6, or two 2×6s and ½-inch plywood nailed together, for a header.

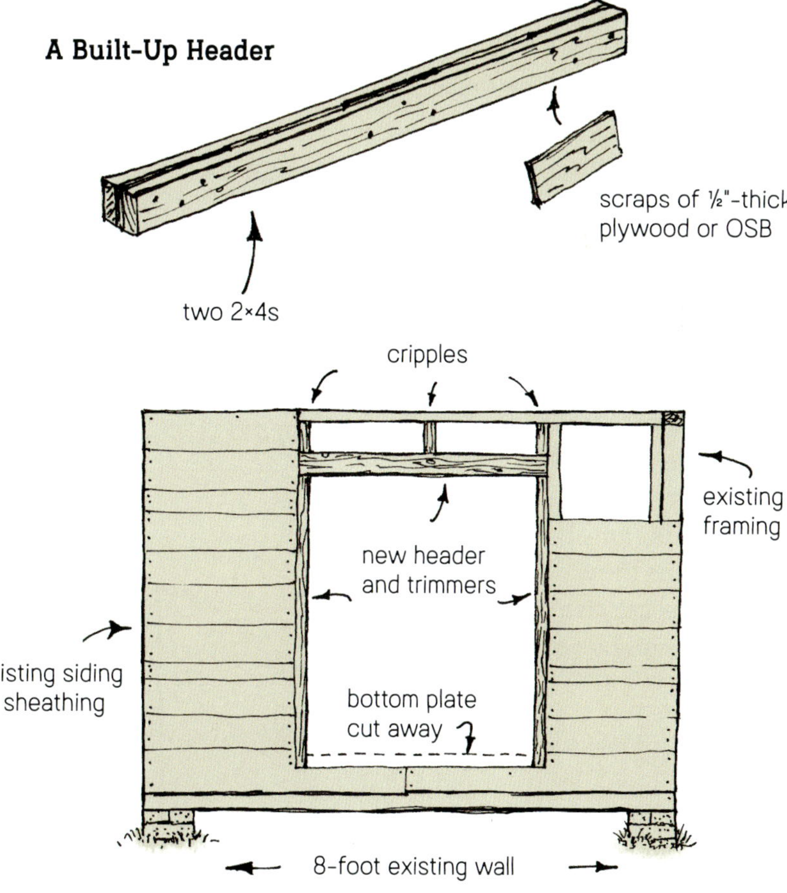

JOINING THE ROOF

Once your doorway is in and the floor is built, you'll need to frame three walls all the same height. Here you will need to think about the roof.

There are several ways you can tie the roof of your addition to the roof of your original structure, depending on how high you want your new walls to be. You can add your new room to the lower end of your structure, with the new roof continuing the slope of the old one. Or you can add your new room to the higher end of your structure. You can join the new roof to the old roof at the peak or just enough below the peak (at least 9 inches) to clear the overhang. Any of these methods allows you to use the same roof-beam design described in Chapter 4.

For our example, we'll join the new roof to the original roof at the peak of the Classic Design. We'll also build the new walls to the same height (5 feet 6 inches) as the walls in the Classic Design, using the same framing methods described in Chapter 4.

Roof Variations for an Addition

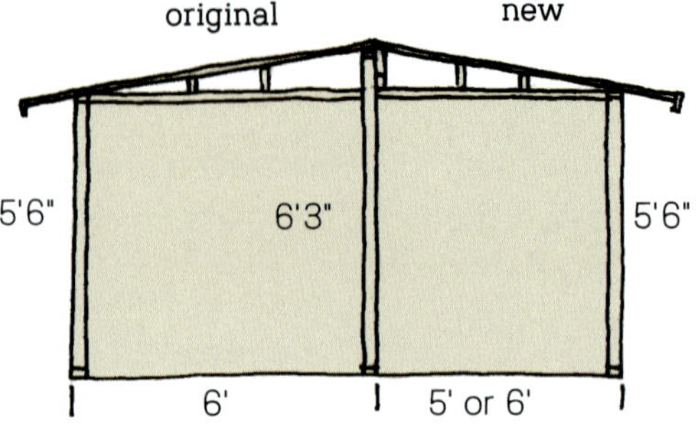

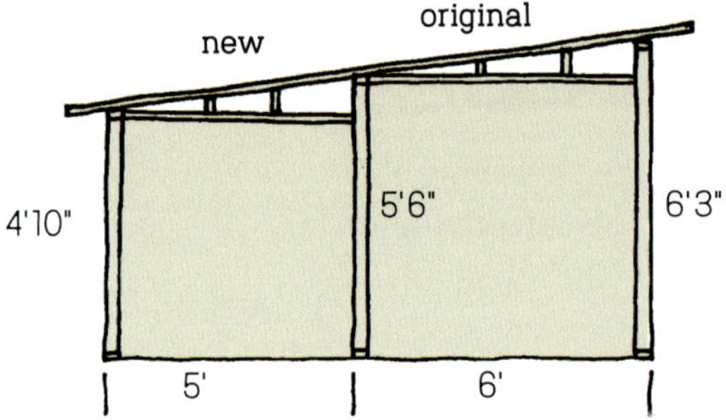

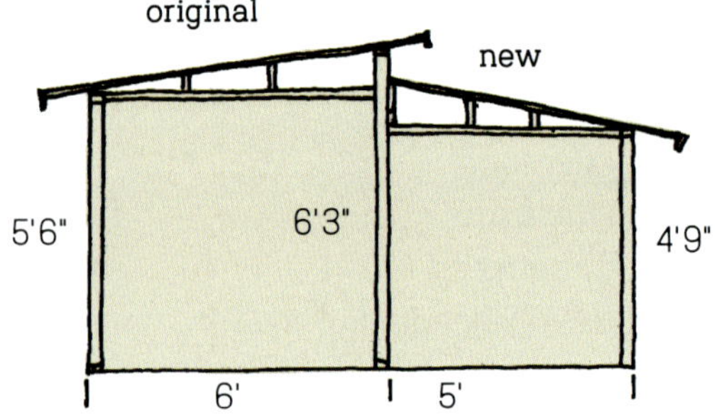

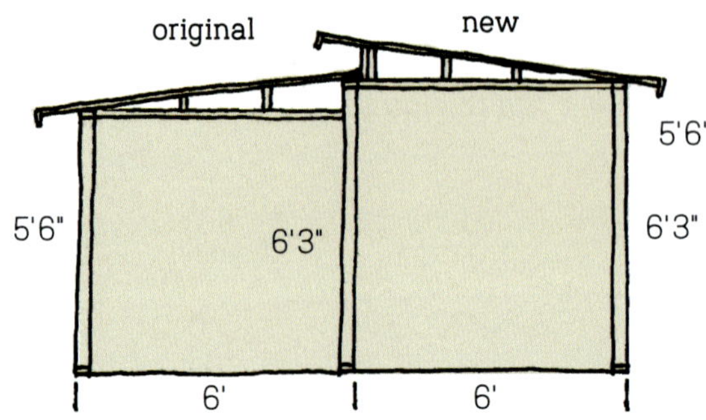

FRAME THE WALLS

Step 1. On your sawhorses, cut the top and bottom plates for the end wall exactly 5 feet (60") long.

Step 2. With both plates set on edge together, measure every 20 inches and draw centerlines for the studs. This measurement (instead of 24 inches as used when framing the original walls) spreads the studs out evenly on this short wall, unless you have a wider door or window. (See Chapter 4 again if you need help to frame in a door or windows.)

Step 3. Cut four more plates for the two side walls at 6 feet long minus the width of the plates of the end wall: 72"–3½" = 68½" long.

Step 4. Set all four of these plates on edge, then measure and draw centerlines every 24 inches for these studs.

Step 5. Cut 12 studs at 63 inches long, which is the height of your wall minus the thickness of your top and bottom plates: 66"–3" = 63".

Step 6. On the ground, lay out the plates with the studs for the three new walls as shown in Chapter 4 (see page 51).

Step 7. Nail each of the three new walls together, using 12d or 16d nails.

Step 8. Stand up the wall frames on your floor one at a time, check them with your level to make sure they are plumb, nail them to your floor, and brace them, as shown in Chapter 4.

Step 9. Nail the corners together, then nail the side wall frames tightly to the side of your original structure (see the drawing below).

Framing the Walls and Trimming the Roof

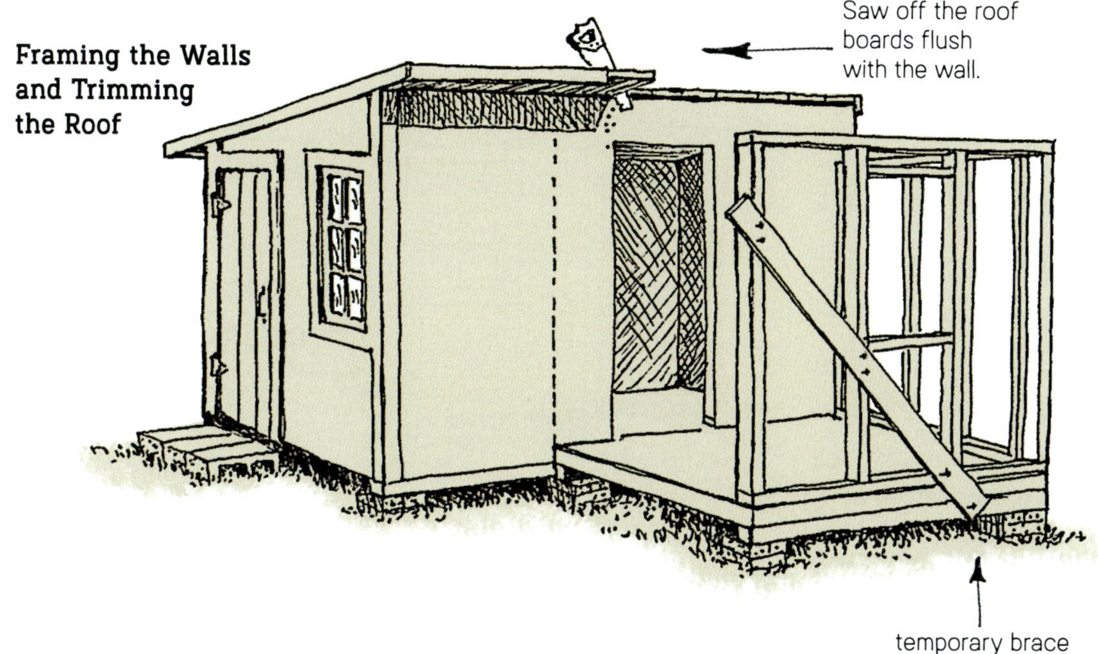

TRIM THE ROOF OVERHANG

Next, saw off part of the original roof overhang so that the new roof and old roof will meet at a peak.

Step 1. Climb up on the roof, or up a stepladder. Bring your saw, hammer, crowbar, and utility knife, along with a straight board and a pencil.

Step 2. Pry off and peel back (as gently as possible) the original roofing. If it tears, don't worry about it; it's easy to repair. Then set your straight board over the roof and draw a cut line where the original roof overhang begins.

Step 3. Saw through the roof boards and roof trim boards until you are just above the new wall frame, as shown in the drawing on page 79.

FRAME THE ROOF

You'll need a stepladder and a helper.

Step 1. On your sawhorses, cut a 2×4 and a 2×6 to 5 feet long.

Step 2. Set them on edge, one on top of the other, on top of your new wall frame against the original structure wall, then nail them both into the wall.

Step 3. Cut another 2×4 and another 2×6 to 5 feet long, and set them across the wall top plates, as shown in the drawing below.

Step 4. Use a straight roof board, or your long level, to determine the proper spacing. The roof board should rest evenly on the new end-wall top plate, the 2×4, the 2×6, and the stacked 2×4 and 2×6 nailed to the original structure.

Step 5. When you have the spacing right, toenail the roof beams to the top of your walls with 6d nails.

FINISHING THE ROOM

Cover the walls and roof of your addition with car siding, boards, OSB, or plywood. Install any windows and doors, as shown in Chapter 5. Add trim boards to the roof, the corners, and around your windows or door, and you are done! Then you can paint or add siding if you want to. (See page 85 for siding details.)

Alternative: Building an Ultra Small Addition

If you only wish to add a small nook or alcove, this option is for you. It uses the same method of construction as the larger addition but is low enough to attach under the existing structure's roof.

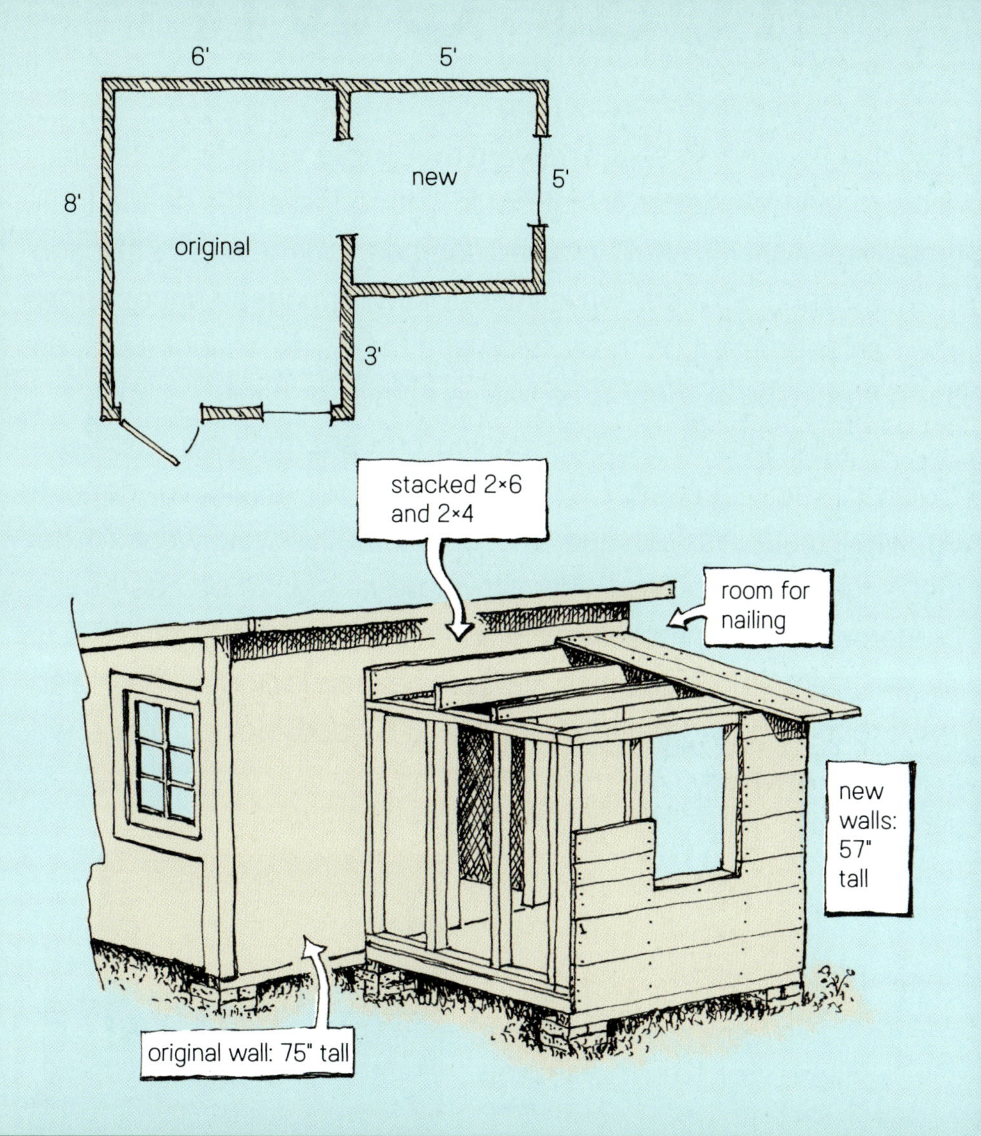

·CHAPTER 7·

Making It Your Own

Now that you've built your backyard dwelling, you might be thinking, "This could look better. What could I do to make this space look more like I want it?" Well, you can upgrade your new construction in a thousand different ways. The following pages suggest how you can decorate it on the outside and the inside, as well as fix up the surrounding outdoor area with a garden or patio.

Fixing up can be a never-ending process for your backyard dwelling, simply because it's so much fun. Your results will depend on what materials you find and how creative you want to be. Once the door is on and your structure is "closed up," as house builders say, you'll probably want to finish the outside first. This is an important step if uptight neighbors have to look at it. You'll also feel proud of how great it will look, and a good-looking space will command attention.

Trimming and Painting the Outside

If you haven't done so already, add some 1×3 or 1×4 trim boards around your windows and doors and on the wall corners. On the corners, extend the trim boards ½ inch *below* the bottom of the sheathing boards covering your structure. This will help later when you put on finish siding. The trim boards cover all those little gaps you might have left and give the place a finished look. Now your space is ready for painting or adding siding of some kind.

The easiest way to decorate the outside is simply to paint it. One gallon of house paint will do wonders and should cover the outside of the Classic Design. If you have more than one paint color, play around with ideas such as painting the trim boards a different color than the walls or painting one side a different color than the other. You're the designer, so the sky's the limit!

Use only water-based paint. Water-based paints are labeled "latex," "latex enamel," or "acrylic-latex." The label might also say "flat," "eggshell," "satin," "semi-gloss," or "gloss." These terms indicate the texture of the paint when it's dry: Flat is really dull, gloss is really shiny, satin and semi-gloss are in between, and eggshell has the texture of, you guessed it, eggshells. As a general rule of thumb, satin or gloss finishes are easier to clean, and flat paints will cover better.

Painting is pretty easy, but it does take some practice to do it well. To start, get a good-quality paintbrush about 2½ to 3 inches wide for painting the walls. Also get a 1½-inch-wide angled brush (called a beveled cutting brush) for painting edges; a beveled brush is especially helpful where two colors meet or you have to paint against the glass on a window.

Painting Tools

2½"- to 3"-wide standard brush for water-based paints

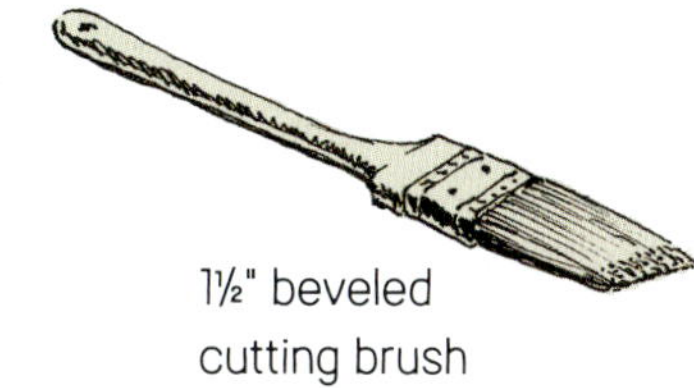

1½" beveled cutting brush

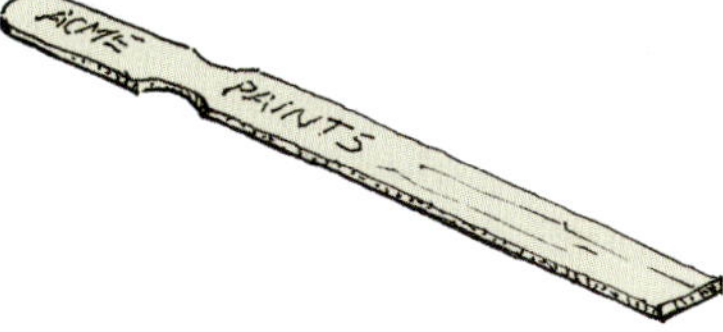

stirring stick

paint-can opener

paint pot

HOW TO PAINT

Step 1. Pry open a can of paint with a flathead screwdriver or a paint-can opener. Stir the paint thoroughly with a clean stick. If the can is more than half full, pour some paint into a smaller paint pot, to about half full or less, then put the lid back on the paint can. A smaller paint pot is easier to handle, and if it spills, you won't lose all your paint.

Step 2. Dip your brush into the pot and smack a little paint off on the inside of the pot so it won't drip.

Step 3. Brush the paint onto the wall with a few long strokes to cover the wood, then work the brush into all the little cracks and holes, especially if the wood is old or rough.

Step 4. Reload your brush and use long strokes over what you just covered. That will help make an even, smooth coat of paint with no drips or bare spots.

Step 5. Keep going! Eventually you'll find the right balance between too much and not enough paint on your brush, and you'll also find the best way to get the paint on the wall without dripping. As with most things, painting will get easier the more you practice.

Step 6. When you're done painting, dump any paint left over back into the can, then brush all the paint remaining on the sides of the paint pot into the paint can. Reseal the can by tapping the lid firmly with your hammer or the side of a 2×4 block.

Step 7. To clean up, fill the paint pot with warm water with a little dish soap, and stick your brush in it for an hour or more to help soak out the paint. Later, rinse the brush in warm water until the water running through it comes out clear—then you'll know your brush is clean. Tap the brush or use a rag to get out the excess water. If the brush came wrapped in a cardboard folder, put it back in the folder to keep the bristles straight. Rinse out the paint pot, and you're done!

Avoid Oil-Based or Industrial Paints

If a well-meaning friend or neighbor offers you oil-based paint or, worse, auto-body or other industrial paints for your dwelling, don't accept. They emit poisonous fumes, and you need toxic solvents to clean your hands and the brushes.

Installing Siding

If you have the itch to keep on building, you might get lucky and find (or be able to buy) some shingles, clapboards, car siding, board-and-batten, or other wood siding for your structure. These are the safest and easiest kinds of siding to put up, requiring only hand tools. Finish siding can make your dwelling look polished and clean.

I don't recommend vinyl siding because special tools are needed to work with it. The same goes for any siding made of aluminum, steel, fiber-cement board, or heavy plywood siding such as T1-11.

Siding Styles for a Backyard Build

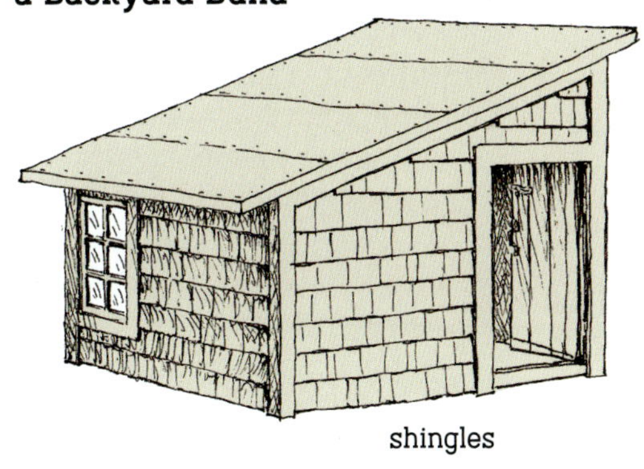

shingles

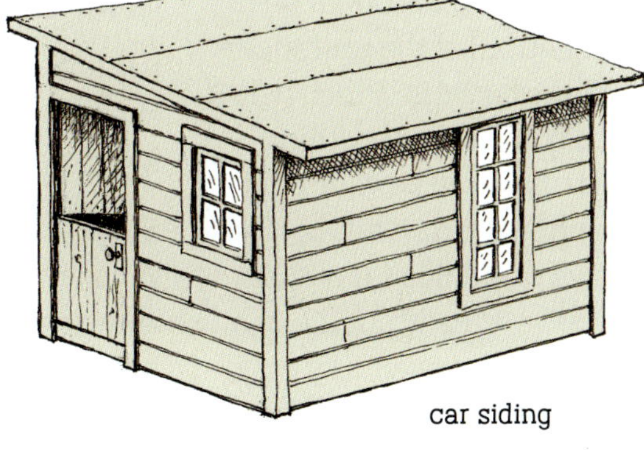

car siding

board-and-batten

clapboard

SIDING WITH SHINGLES

Shingles have been used to cover houses for hundreds of years. They are thin, tapered strips of wood (usually cedar) about 16 inches long and 2 to 10 inches wide. The thicker bottom end of a shingle is called the butt. Shingles are graded like lumber, from no. 1 to no. 3, and are sold in bundles. I recommend no. 3, or "backup," shingles for their low cost and wide availability. Shingles are quite easy to put on, and after a while you'll understand the logic of how they shed water. You will need your hammer, a 1-pound box of 3d galvanized box nails, a utility knife, and a handsaw.

HOW TO SHINGLE

Step 1. To start, temporarily nail or prop up a straight board at the bottom of the wall so the top of the board is ½ inch below the bottom of the sheathing boards. This will be your guide for nailing the first two courses, or rows, of shingles. (I'm assuming you've already put the trim boards on the corners by now, because the shingles will be nailed up against the trim board edges.)

Step 2. Load up your nail apron with 3d nails.

Step 3. Grab a wide shingle and set it on your guide board and against the corner trim board.

Step 4. Drive a nail about 7 inches above the bottom of the shingle and ¾ inch in from each side edge of the shingle. No matter how wide the shingle is, use only two nails per shingle.

Housewrap

Most builders put tar paper or, nowadays, a special white plastic called housewrap on house walls before nailing on the siding, which helps keep the inside dry. Shingles, clapboards, or other siding will do a good job by themselves, especially for a backyard dwelling, so it's up to you. If you have tar paper (not the mineral-coated kind) left over from your roof, save money and just use that.

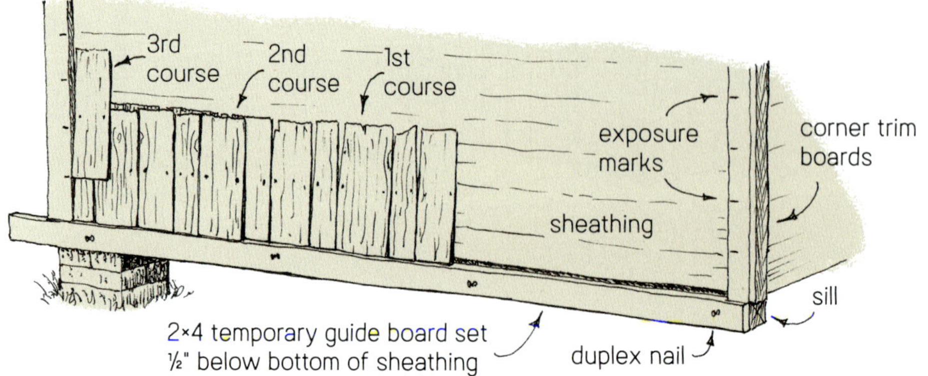

The bottom row of shingles is doubled to protect against water infiltration.

Step 5. Add more shingles, mixing narrow and wide ones in a random pattern, as shown in the drawing on the facing page.

Step 6. As you nail the second course of shingles over the first one, stagger the shingles by keeping the vertical joints or cracks between shingles at least 1 inch away from those on the row of shingles beneath.

Step 7. Continue nailing both rows all the way to the other end of the wall.

Step 8. When you get to the far end, set the last shingle against the trim board and mark its top and bottom with your pencil so you can cut it to fit.

Step 9. Get out your utility knife. Set the edge of a straight piece of wood on the shingle at the two marks, and pull your knife along the piece of wood to cut the shingle. This works well until you hit a knot; if this happens, use your saw to cut the shingle.

Step 10. Measure up from the bottom of both trim boards, making pencil marks every 6 inches (6", 12", 18", 24", and so on). This marks the exposure of your shingles by showing you where the bottom of every row of shingles will be. Builders call this "6 inches to the weather."

Step 11. Start the next row of shingles at the corner board. Line up the bottom of a shingle with the mark you made on the corner board, check the vertical edge to make sure it is not over the crack between the shingles below, and nail it in. Add more shingles in the same way.

Keeping Courses Straight

To keep your rows of shingles straight, there are two methods:

Guide Block. For short walls and most backyard dwellings, you can use a guide block. This is an easy-to-make piece of ¾-inch-thick wood with a 6-inch notch cut into one side. Simply hold the block so the notch is hooked on a shingle you have already nailed in, then set the shingle you want to nail on the top of the block. Remove the block and nail in your shingle.

Chalk Line. For long walls, you can use a chalk line, which is a long string with a little hook at the end that is kept inside a "chalk box" filled with bright blue or red chalk. Hook the string to a small nail (or have a helper hold it) at the exposure marks on both ends of the shingle row and pull it tight. Then pull the string straight out from the middle of the wall and let it snap back; the chalk leaves a line that is your guide for aligning the bottoms of the shingles.

HOW TO SHINGLE AROUND A WINDOW

When you get to a window, you'll have to trim the shingles to fit around it.

Step 1. Set the first shingle to be cut over the window trim and hold it steady while you mark where it needs to be cut.

Step 2. On your sawhorse, first cut across the grain of the shingle with your saw. Then switch to a utility knife to cut with the grain, as before. This lessens the chance of splitting the shingle.

Step 3. Nail the shingle in place.

Step 4. Cut the remaining short shingles that go under the window to the same length with your saw. If these are shorter than 4 inches, use a trim board instead. The same method applies when you get to the top of your wall.

If you goof while installing a shingle (it happens!), use your crowbar to pry it off. Pound the short leg of the crowbar under the shingle at the nailhead and then pry. You might lose part of the shingle (they split easily), but the middle might still be usable.

If you goof, pry up the nails and try again.

SIDING WITH CLAPBOARD OR BEVELED SIDING

Clapboard, also called beveled siding, is somewhat easier and faster to put on than shingles. Clapboards are set against the corner trim boards, just as shingles are. They are usually 6 or 8 inches wide, about ½ inch thick at the butt (the bottom edge), and beveled, or tapered, to about ⅛ inch thick at the top edge. The 6-inch-wide clapboards are best for the scale of a backyard dwelling.

The standard exposure of wood to weather for 6-inch-wide clapboard is 4½ inches. Some clapboard styles have a special ledge, or "rabbet," in the back that guides your exposure for you. If not, make a guide block that is set for a 4½-inch exposure, as described on page 90.

A Note on Scale

In architecture, scale is the idea of using right-sized objects or patterns to fit a certain size of building. A smaller structure needs smaller doors and windows, as well as small-patterned finishes like shingles and skinny clapboards, to give it the right scale.

HOW TO PUT UP CLAPBOARDS

Putting up clapboards takes a lot of nailing, so put on your nail apron and load it up with 6d galvanized box nails.

Step 1. Set a long, straight clapboard at the bottom of the sheathing. Have an assistant help hold it there. If one end overlaps the corner trim board, use a pencil to mark the point where the clapboard meets the inside edge of the trim board. Then take the clapboard to your sawhorses, draw a cut line at the mark using your square, and saw off the scrap. (Alternatively, instead of using a pencil, you can draw, or score, the cut line with your utility knife to help guide your saw for a clean cut.)

Step 2. Bring the clapboard back to the wall and set it against the sheathing so that it overhangs the bottom of the sheathing by ½ inch. Have your assistant help you hold it while you tack it in place with a nail or two.

Step 3. Drive a nail every 16 to 24 inches along the board, about 1 inch up from the butt. If you can, drive the nails into the studs beneath the sheathing, so they won't poke through on the inside. When you get to the end of a clapboard, drive a nail 1 to 2 inches from the end. And as you work, keep each clapboard end at least 6 inches away from a joint in any adjoining course, or row.

Step 4. When you've finished the first course of clapboard, it's time to mark the location of the remaining courses on the corner trim boards so you can align the clapboards evenly. For 6-inch-wide clapboards, measure up from the bottom of the corner trim boards and mark at 4½ inches, 9 inches, 13½ inches, and so on, to set the bottom edge (or butt) of the clapboards for each course. Then make a guide block similar to the one for shingles but set for 4½ inches of exposure instead of 6 inches (see page 90).

Step 5. Set each clapboard to the marks on the corner trim boards, and, using your guide block, nail it in.

Plain or Rabbeted Siding

6" wide

4½" exposure

4½"

6d nail here

Plain Bevel

starter strip

no guide block needed

5" wide

4½" exposure

4½"

6d nail here

Rabbeted Bevel

no starter strip

A Starter Strip

For a professional look with clapboards, nail on a starter strip at the bottom of the wall first. The starter strip will hold the first clapboard at the same angle as the clapboards above it. Cut a 1½-inch-wide strip off a clapboard's upper edge and nail that with some 3d galvanized box nails along the very bottom of the sheathing (not ½ inch below it). Then set the first clapboard over the strip and ½ inch below the sheathing, and nail it in.

Putting Up Clapboards

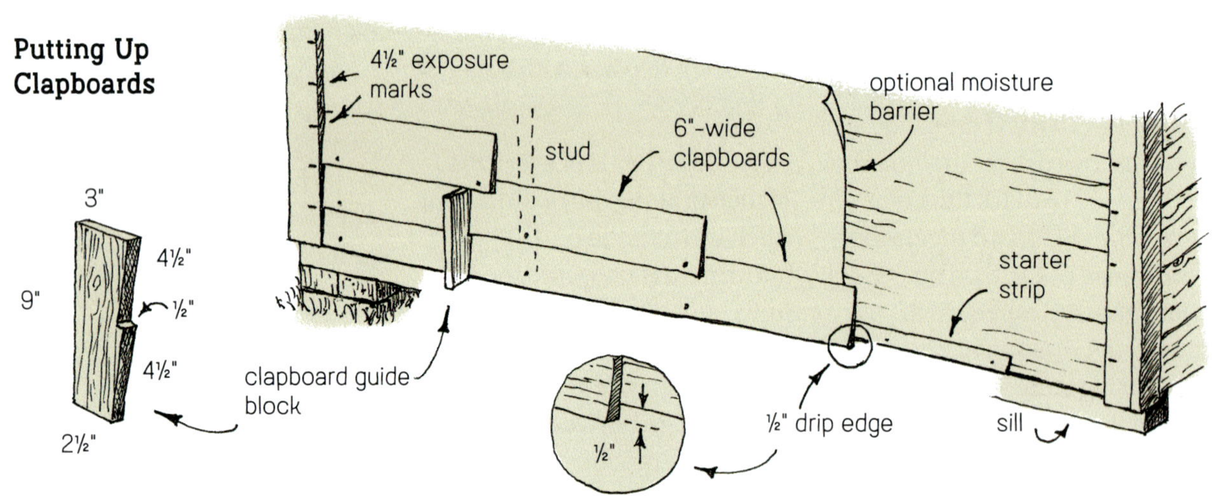

HOW TO INSTALL CLAPBOARD AROUND A WINDOW

When you get to the bottom of a window, you'll have to cut away some wood to fit around the window trim boards. This isn't difficult, but it does take some patience.

Step 1. Set the clapboard over the window trim, as shown below, and use a pencil to mark the part that needs to be cut out.

Step 2. On your sawhorses, saw the vertical edges of the notch to the corners.

Step 3. To cut along the horizontal line, get your hammer and a chisel. Use a clamp or have a helper hold down the clapboard. Hold the chisel's cutting edge on the line, with the flat side to the good part of your board, and hit it hard enough with your hammer to cut through the wood. Keep chiseling along the horizontal line until the scrap breaks off. Watch the grain of the wood so your cut doesn't break into the good part. If it starts to do that, go to the other end and chisel the other way.

Step 4. If you are cutting out a long piece, use your handsaw to finish the job once you have some room. You can also use your utility knife and a long steel ruler to cut through it if the wood is thin enough and there are no knots.

Cut Out the Scrap

HOW TO INSTALL CLAPBOARD AT THE TOP OF A WALL

When you get to the top of the wall, you can simply stop. If there is a gap between the last clapboard and the roof overhang, cover it with a trim board. The remaining clapboard left over from your starter strip might fit up there if you're lucky. If the top of the wall is slanted, you'll have to cut the clapboard to fit.

Step 1. Mark the ends of the clapboards where they overlap the trim board at the roof.

Step 2. Connect the marks with a cut line and saw off the scrap.

If the roof overhangs too far to mark the clapboard, measure from the point where the clapboard hits the overhang to the same point of the clapboard below. Then draw a line between the marks and saw off the scrap. If you need to carve off a bit more wood, try using your utility knife.

Clapboarding Against a Slanted Roofline

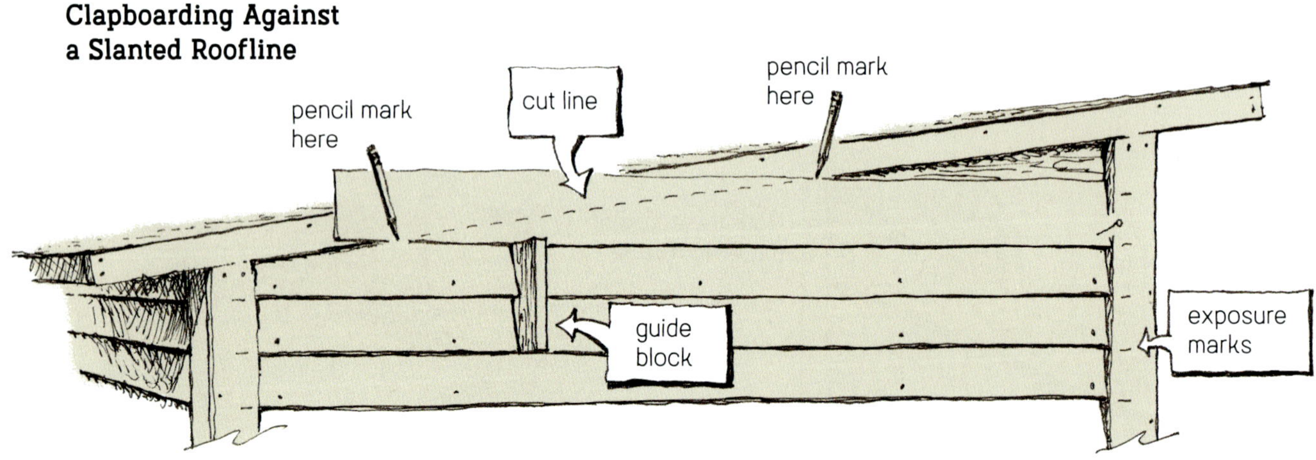

SIDING WITH CAR SIDING

Named for their original role lining the inside walls of railway cars, pine car siding boards are ¾ inch thick and 6 or 8 inches wide, and they are tongue-and-grooved for a tight fit. Car siding boards usually have V-grooves on one side and are smooth and flat on the other.

As I mentioned in Chapter 5, car siding is strong enough to use as sheathing and finish siding, and it is relatively inexpensive. If you have to buy sheathing and you like the look of car siding as a finish siding, then by all means, save time and money by just using car siding! Use it smooth side up for a strong floor and roof as well.

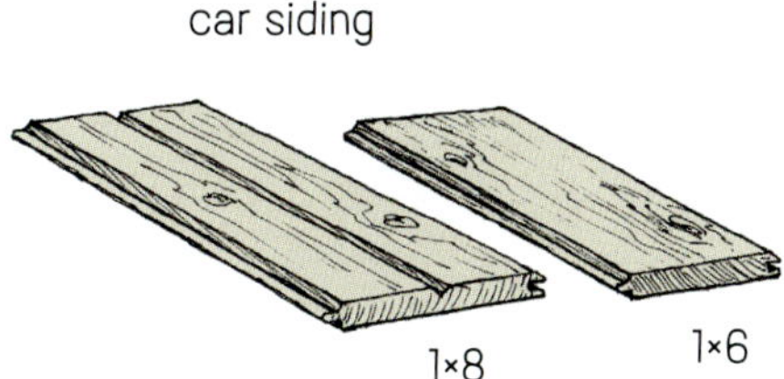

HOW TO INSTALL CAR SIDING

To install car siding as sheathing and finish siding, you'll cover one wall at a time while leaving the brace boards up on the other walls. For strong walls, measure and cut each board so its ends are set over studs.

Step 1. Start the first board ½ inch below the bottom of the floor joists for all the walls (see the drawing on page 94). Making sure the tongue is at the top of the board, have an assistant help hold the board, and nail it in with 6d or 8d galvanized box nails. Use at least two nails at each end of the board and two nails over each stud.

Step 2. Fit the groove of the second board firmly over the tongue of the first one. If they don't join tightly, look in the groove for anything stuck in there. If they still resist joining, use a scrap of wood as a hammer block and pound the boards together (see the drawing on page 94). Then nail the second board in place.

Step 3. Continue nailing on boards as described above. When you get to a door or window opening, measure and cut the boards carefully so you can nail them tightly and neatly against the framing. You can be less careful on the corners because the corner trim boards will cover up any small gaps.

Step 4. When you get to the top of the wall, measure the space needed for the last board, mark the board with your pencil and a long straightedge (a long, straight board will do), and then rip it with your saw. As a reminder, to "rip" means to saw a board along its length. This can take a while, so use clamps or have someone hold the board steady and also help you saw it.

Step 5. When all the car siding is nailed on, install the trim boards on the corners and around the windows as you would with rough sheathing.

Installing Car Siding

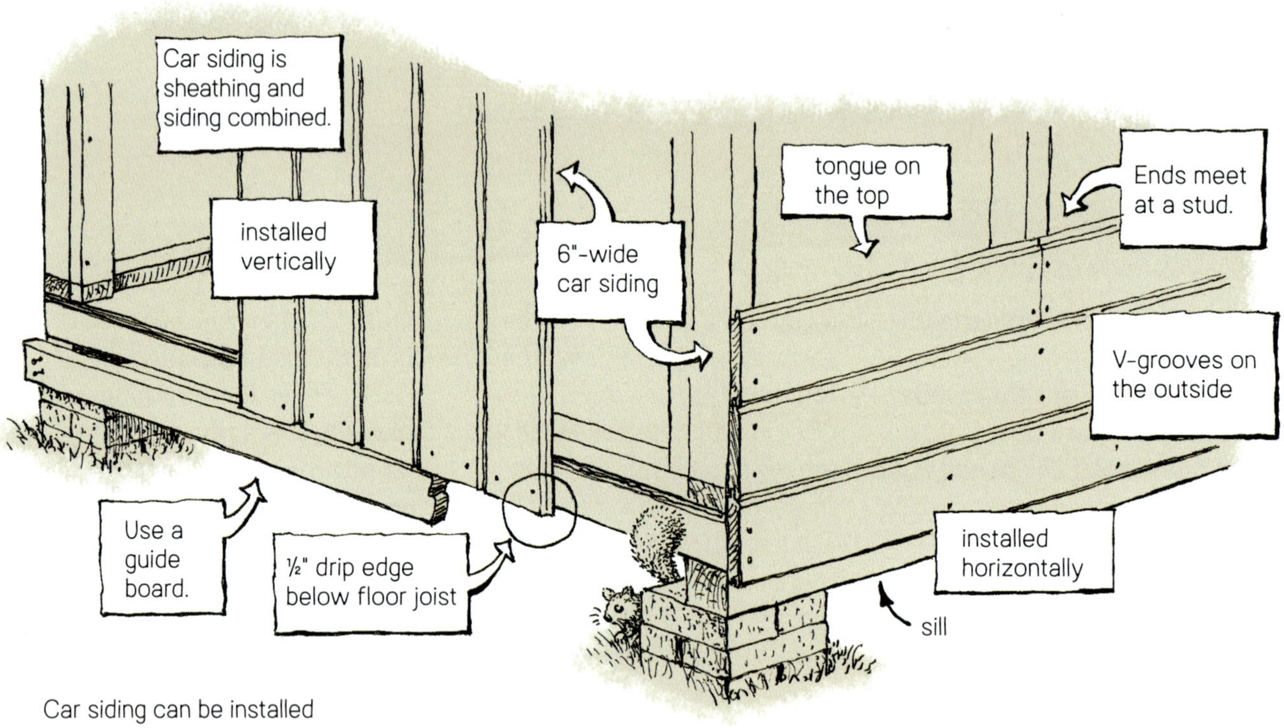

Car siding is sheathing and siding combined.
installed vertically
6"-wide car siding
tongue on the top
Ends meet at a stud.
V-grooves on the outside
Use a guide board.
½" drip edge below floor joist
installed horizontally
sill

Car siding can be installed vertically or horizontally.

Joining a Stubborn Board
grooved scrap

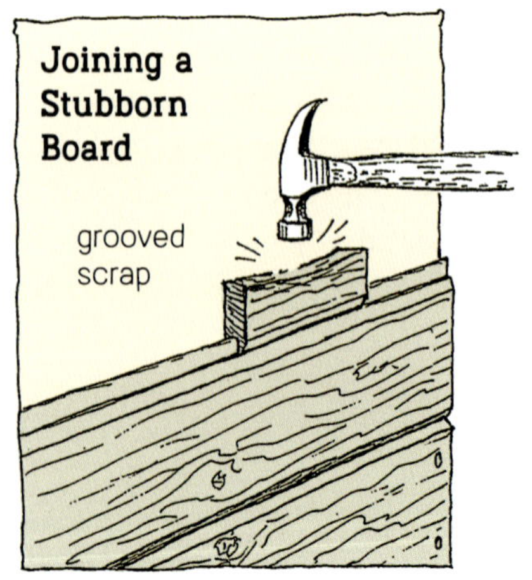

SIDING WITH BOARD-AND-BATTEN

Board-and-batten, or board-and-bat as it's sometimes called, is another style of siding that can double as sheathing. This method uses plain, wide, ¾-inch-thick boards installed vertically. The joints between the boards are then covered with narrow boards or strips about 1½ inch wide, called battens or bats.

Old shed, barn, and fence boards—or any boards wider than 5 inches—will work well for board-and-batten. For battens, use 1×2 furring strips.

HOW TO INSTALL BOARD-AND-BATTEN

Cut the boards and battens long enough so they reach from the top plate to ½ inch below the bottom of the floor joists. Save any short boards for under the windows or over the door. For slanted walls, cut the tops of the boards at the same angle as the roof. To do this, hold the board in place and mark the angled edge of the roofline on the board with your pencil. Use your square to draw the cut line, then saw off the scrap.

Step 1. Add a 2×4 block horizontally between each pair of studs, halfway between the top and bottom plates (see page 96). These will strengthen the board-and-batten wall.

Step 2. Starting at a corner, nail the boards to the middle blocks and the plates, using three 8d galvanized box nails at each end and two in the middle. If a board rests over a stud, nail it into the stud every 16 inches or so. Use your level to check the boards once in a while to make sure they are plumb; they can get crooked.

Step 3. When the walls are covered with boards, nail on all of your trim boards on the corners and around the door and windows *before* you nail on the battens. You can use extra battens as trim boards if they are wide enough.

Step 4. Nail on a top trim board where the wall meets the underside of the roof.

Step 5. Nail on all the vertical battens with 4d galvanized box nails, spacing the nails every 16 inches.

Board-and-Batten Siding: Nailing on Boards

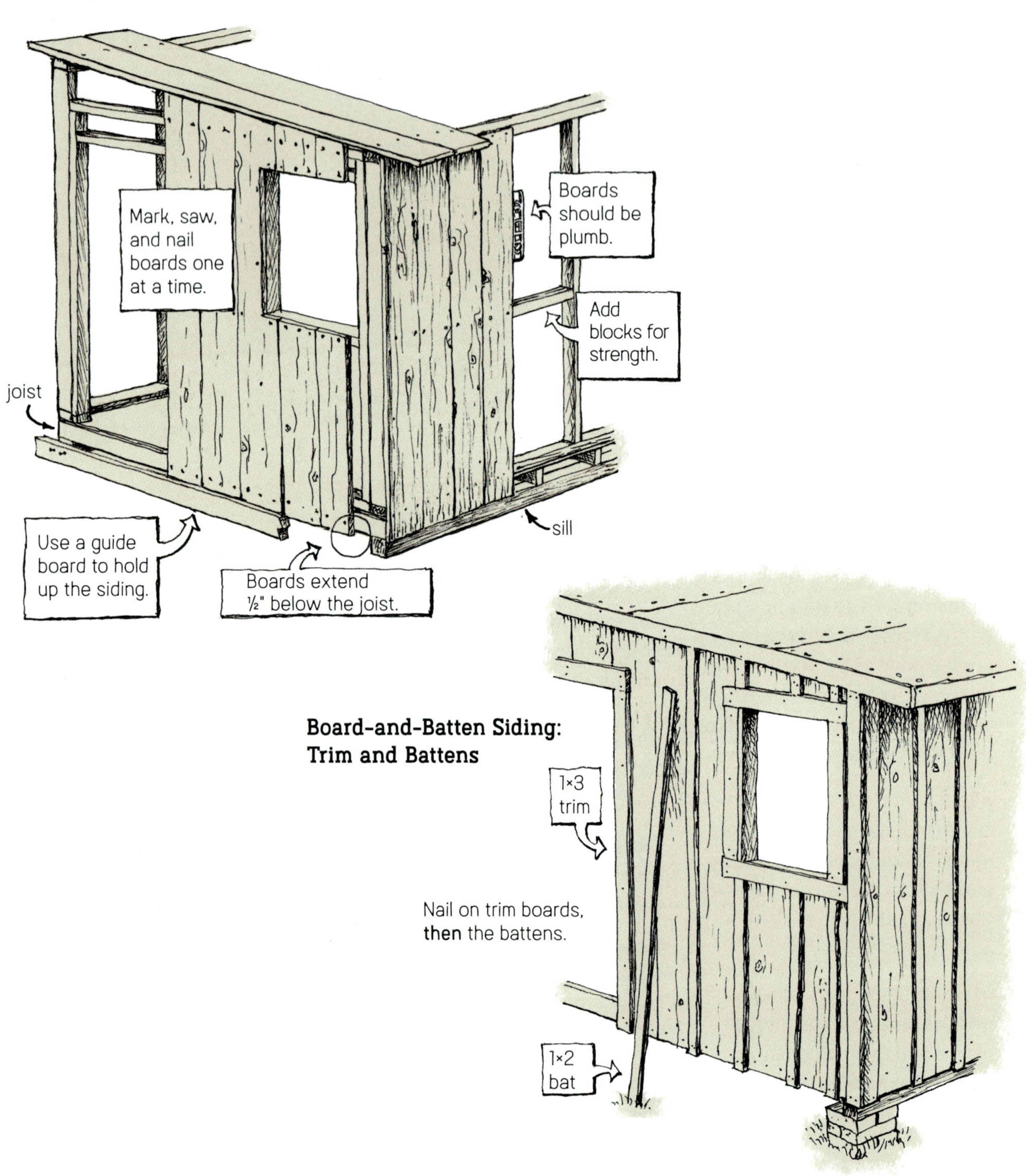

Board-and-Batten Siding: Trim and Battens

Nail on trim boards, **then** the battens.

Fixing Up the Inside

As you spruce up the outside, you're probably thinking of ways to fix up the inside as well. Your results will again depend on what materials you'll find and also how houselike you want your backyard dwelling to be. You can be creative with wallpaper, curtains, pictures, posters, signs, paneling, and, of course, paint. Garage sales are great places to find tables, cabinets, chairs, knickknack shelves, pictures, fancy light fixtures, and whatever else might work.

PANELING THE INSIDE WALLS

If you want to finish the inside of your space with some kind of paneling, you will need to add some 2×4s, called nailers, to the inside corners so that you have something to nail the paneling to. Find or cut some 2×4 blocks about 12 inches long and nail them against the corner studs, as shown in the drawing at right. Then cut a new stud the same length as the one in the corner, and nail that in. Now you have a surface on which to nail the outer edge of your paneling.

Nailer Studs for Inside Corners

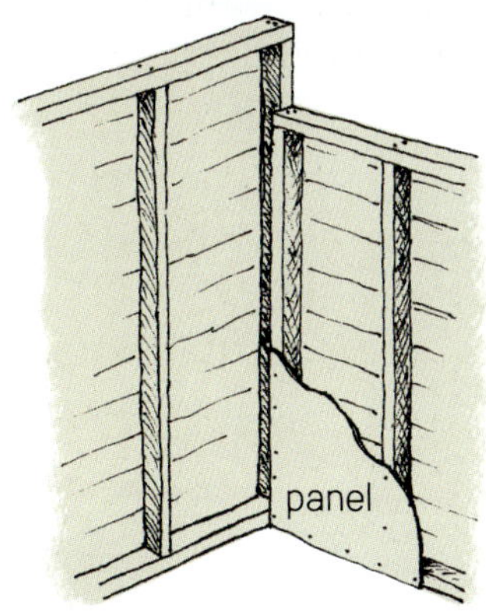

At an inside corner, only one wall has accessible framing to nail the paneling to.

So, add 2×4 blocks . . .

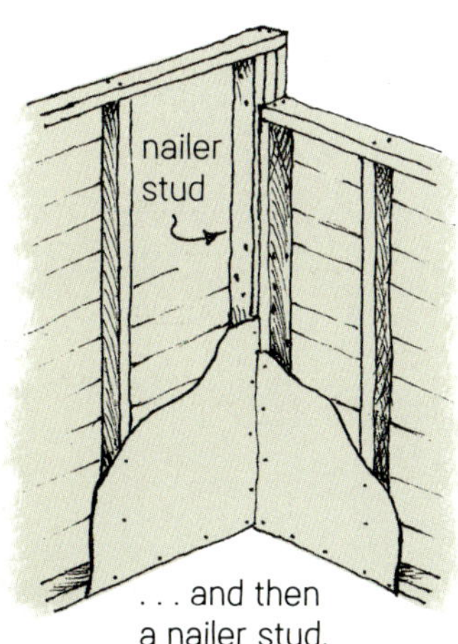

. . . and then a nailer stud.

Paneling Options

For paneling your walls, I recommend using anything made of wood. Thin boards that fit together tongue-and-groove style, called beadboard, are good for paneling, but any smooth boards will do. Many styles of interior wall paneling are made of thin fiberboard or plywood covered with printed designs or painted, but this stuff can be expensive as well as difficult to put up. A far cheaper alternative is ¼-inch-thick OSB, whose finished side is smooth enough to paint or apply wallpaper.

You can also use drywall, otherwise known as wallboard, gypsum board, or Sheetrock, but it is heavy to lift, requires plastering skills to finish, and can get moldy if your dwelling is in a shady or damp place. If you still want to use drywall, see page 162.

Part 3

THE ENHANCED DESIGN

In this section, I'll demonstrate how to build an 8 × 15-foot permanent structure—an affordable project that anyone who can swing a hammer can build with their own hands. Once you see how it all goes together, you can use this design, change the measurements to suit your needs, or build something altogether different.

Many of the building methods used for the Classic Design described in Part 2 are the same ones used for an enhanced structure. If you build the Classic Design first, you'll be all set to tackle this slightly more advanced project.

At only 120 square feet on the ground, this 8 × 15-foot retreat is adaptable to many uses, while the shape fits in a relatively tight space. The gable roof allows room for a loft and gives the building an elegant look.

This dwelling is also portable: The self-bracing plywood construction will support this little house should it be lifted or built onto a flatbed or trailer chassis. With a road height of less than 14 feet, it is low enough to pass under bridges.

·CHAPTER 8·

Planning

You may have seen a potting shed, guest cottage, or studio you would love to recreate in your own yard. Or maybe you have a vision for a private get-away-from-it-all space of your very own. Whether you're starting with this more ambitious plan or you've already built a smaller structure, take a good look at your property and envision the space you would like to have there. (See Finding the Best Building Site on page 27.) Don't be afraid to dream big!

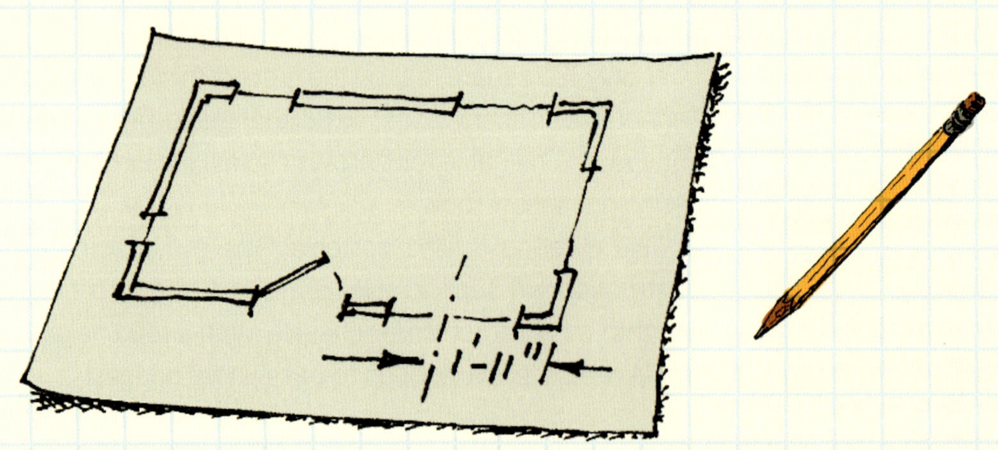

Designing It

There are many ways you can arrange your dwelling. Sleeping, eating, working, and relaxing can all happen in surprisingly tight quarters. You could start by building a Zen-like empty room and decide how you'll use it later. Or you may want to figure out every detail beforehand. It's your choice! This is where your dream place becomes your real place.

First make a list (see Make a List of What You Want on page 27): Write down what you would like to do in your new building, then list all the things you would need to accomplish this. For example, if you want to paint landscapes, you'll need room for your easel, supplies, and other equipment; a place to rest or to stand back and evaluate your progress; and good lighting. You might want to look in magazines and design books for art studio ideas.

Also, think about how your structure will relate to the site you've chosen. Will you want to have a garden outside the door, or a small patio or a hot tub for relaxing? Does the space outside the structure offer good sunlight for a garden or privacy for a hot tub?

Next, begin to sketch out your ideas. With your list handy, start with a rectangle (with our 8 × 15-foot project size in mind), and label the different areas of the rectangle with "painting area," "storage," "computer/printer," and so on, depending on what you want to do. Think about where you want to enter your dwelling and where to let in light and air. What will the inside look like? The outside? When sketching, feel free to add as much room as you'd like, but keep in mind that you are going to do the building.

Once you've established the size of your structure and how you'll use it, draw more detailed sketches of your floor plan and side views. If you've never done this before, the exercise can be challenging, but try to get your ideas down as clearly as possible. And ignore any negative criticism, be it from your family or friends or even yourself . . . this is your place, after all, and you can build it any way you want it!

Our example project, shown below, is a hideaway with a loft and room for a studio workspace and/or living quarters. If this layout doesn't appeal to you, take a look at the other plans on the facing page, or feel free to sketch out your own.

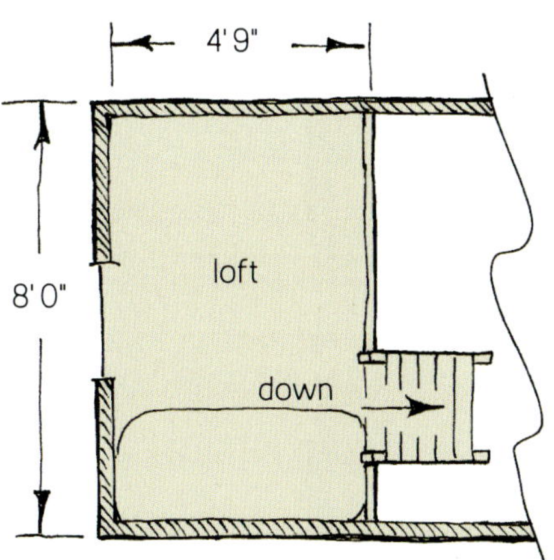

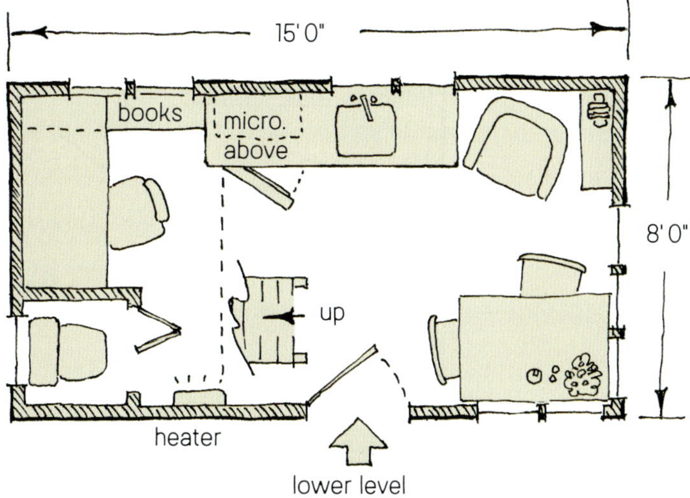

Sample Hideaway Plans

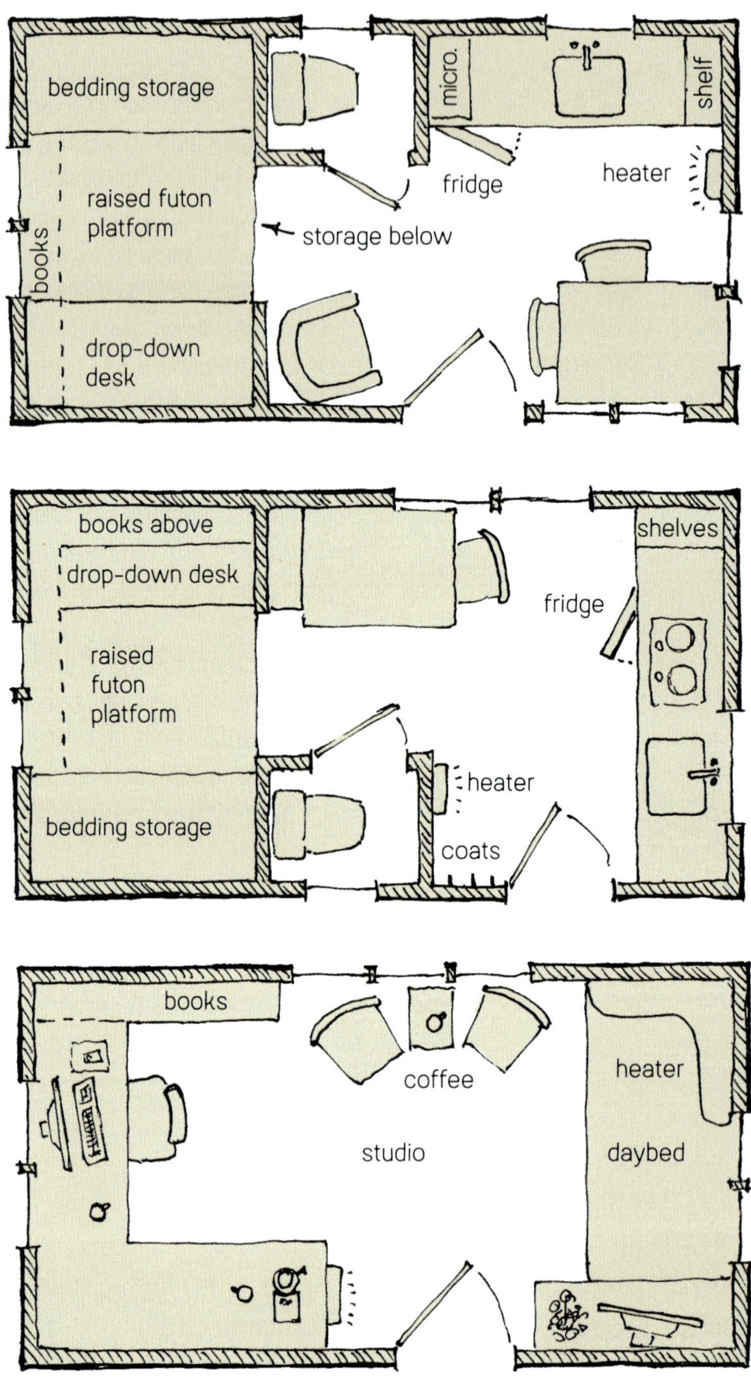

In most places, you don't need a building permit to build a structure that's 120 square feet or less. Nevertheless, be sure to check with your county, city, or HOA before you begin.

Start with Windows and a Door

Once you have a good idea of what you want to build, start your planning by thinking about windows and a door. As you draw your floor plan, it is wise to know the exact dimensions of all the openings. You might even want to have the door and windows on hand before you begin drawing.

You can install used windows or new ones; their size will be up to you and what is available. For such a small space, I believe single-glazed windows are adequate for most climates outside Alaska. Have at least one openable window near each end of your shelter for good cross-ventilation.

A good door for your structure would be at least 32 inches wide, exterior grade, and already on its hinges in a frame or doorjamb. These "prehung" doors are offered in many styles at home centers and lumberyards. Some recycled building-supply stores sell salvaged prehung doors as well, which might have just the right look and are usually far more economical than new ones.

In this example project, we'll use some recycled windows that are 18 inches wide by 36 inches tall, and then we'll gang them up to make 36 × 36-inch and 36 × 54-inch windows. Above the loft, we'll use a 20 × 25-inch barn sash window, which is available in many large home centers. This can be ganged up as well. For the tiny toilet room/closet, we'll use a recycled window that is only 12 × 24 inches. The door here will be 30 inches wide by 80 inches tall.

Door and Window Locations

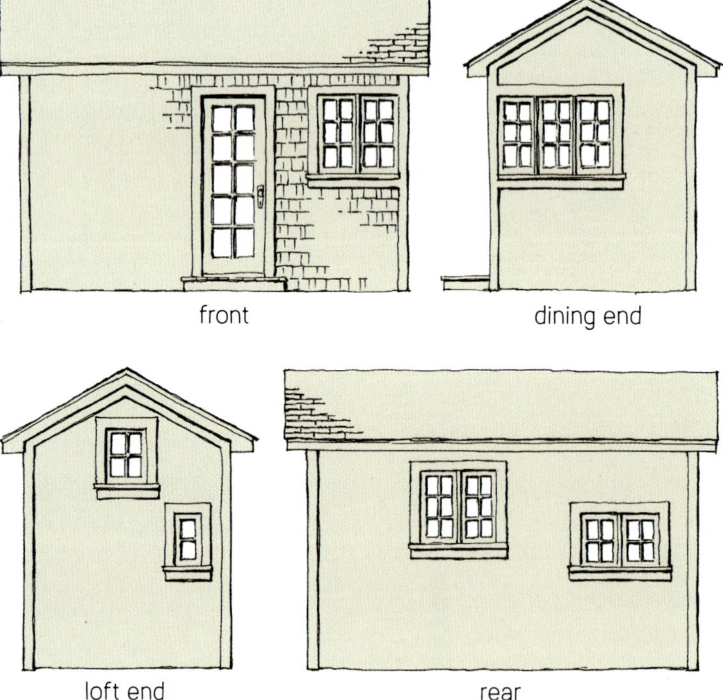

Drawing a Floor Plan

Once you've decided what you want in your dwelling and have determined the window and door sizes, you can finally draw a dimensional floor plan to scale. Yes, this might take you back to high school drafting class, but a simple floor plan with dimensions is relatively easy to produce and quite crucial—you'll be glad you did it!

When you have the windows and the door you want, list each one and its dimensions on your floor plan. Put a letter beside each window on the plan, then use the same letter to identify the corresponding window measurements on your list. This list, called a schedule, will tell you how large to build each window and door opening.

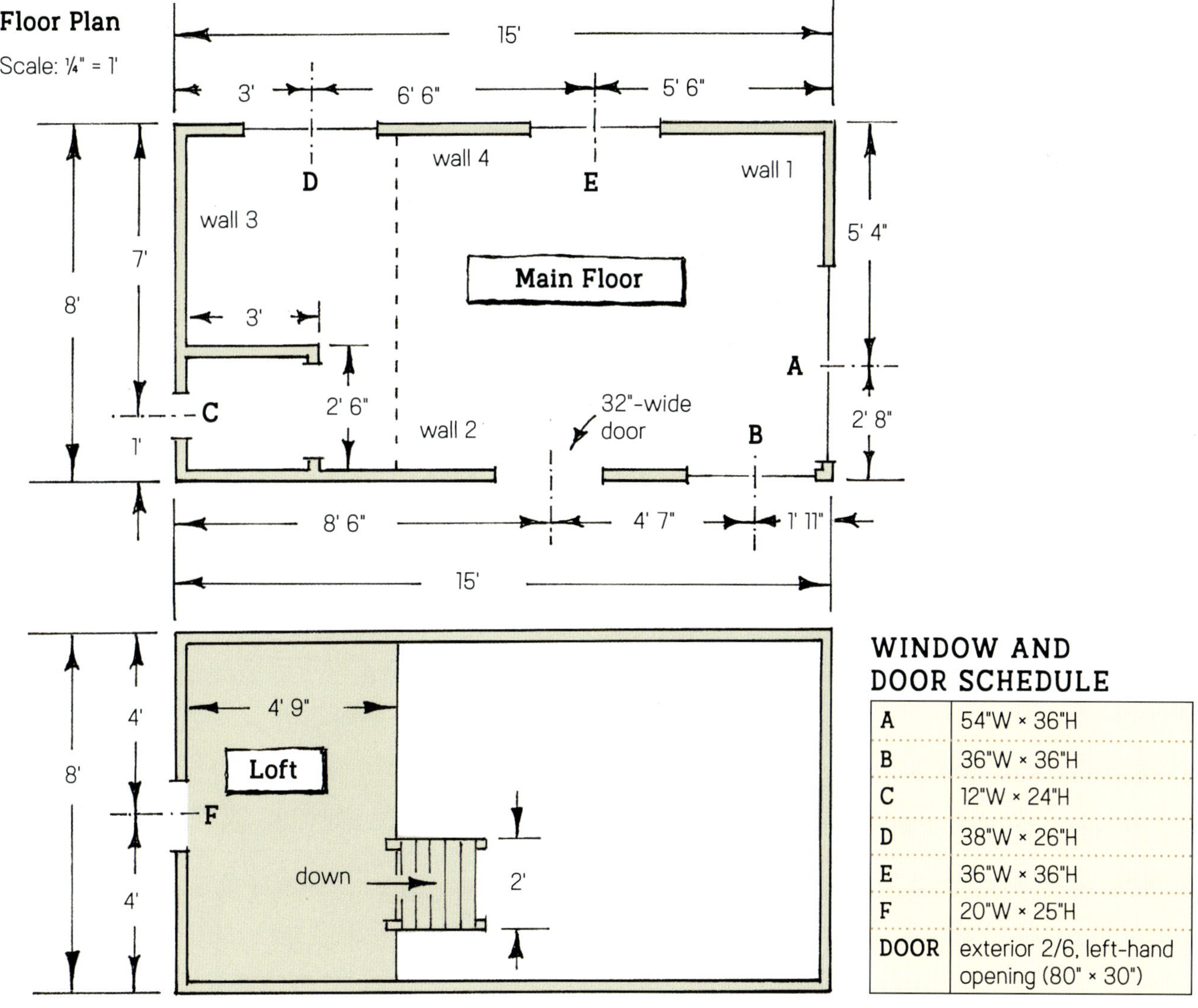

WINDOW AND DOOR SCHEDULE

A	54"W × 36"H
B	36"W × 36"H
C	12"W × 24"H
D	38"W × 26"H
E	36"W × 36"H
F	20"W × 25"H
DOOR	exterior 2/6, left-hand opening (80" × 30")

One way to draw your plan is to use graph paper and decide how many inches each little square will be (the scale). Another way is to get an architect's scale at any drafting or office-supply store. With this three-sided ruler, you can choose a scale for your plan. With "scale," you are saying that ¼ inch, ½ inch, or some other measurement on your paper equals 1 foot in reality. I like to use the ¼-inch scale, but the ½-inch scale might be easier to work with for beginners. To use this scale, look for the ½ number at the end of the architect's scale. You'll see that the feet begin at 0 and run the length of the ruler. The inches are the tiny marks that begin on the other side of the 0. This way, you can measure out the feet and then add the inches, all from the 0 mark (see the drawing below).

To draw your floor plan, get a plastic drafting triangle, a T-square, tape, drawing paper, and a sharp pencil, along with your architect's scale. Find a table with a straight side for your T-square. Tape down a sheet of paper. From your sketch, measure out the length and width of the floor, then draw the walls, as shown on page 105. Look at your rough sketch to locate where you want your window and door openings on the plan. Then with your scale, find the *centers* of the openings on your plan and measure out to their edges, depending on how wide they are. This plan, along with your outside wall sketches, will provide enough information for you to start your building. Make a copy to take out to your site.

How to Use an Architect's Scale

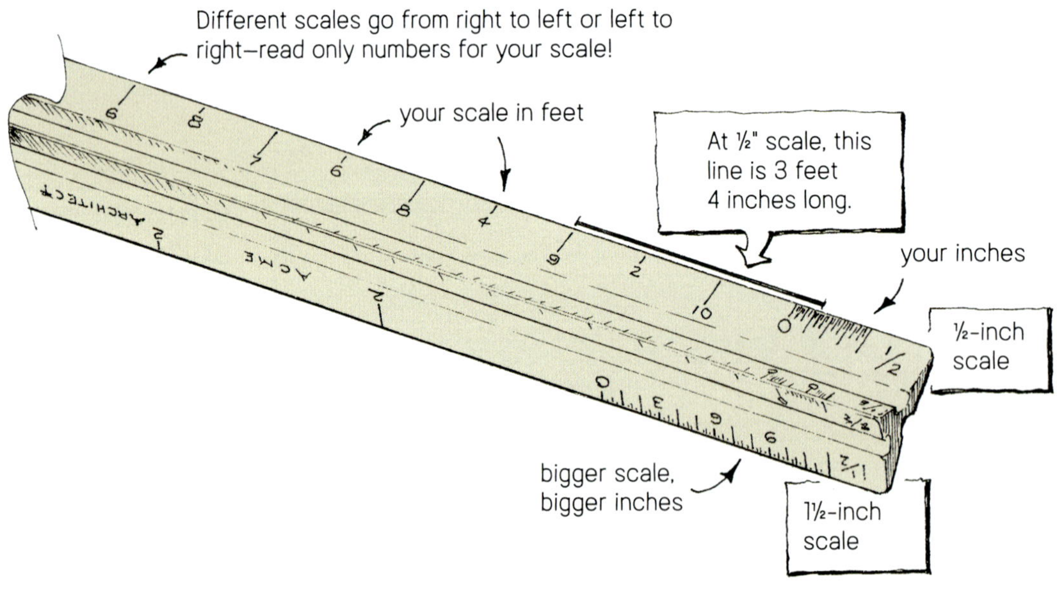

Gathering Materials

Now you are ready to get the lumber and other supplies to build your structure. Each chapter that follows includes a materials list for a specific phase of your project. Of course, you can gather everything at once, but if your supplier is nearby, you might want to get small loads as you build.

In general, it's not difficult to estimate materials. In most cases, you can simply count up the pieces you need, depending on the area of your floor, roof, and walls. It is wise to add a few more pieces to your total, as a cushion; there's always that extra board or two you never thought of in the plans. To estimate wall studs that are set at 16 inches on center, calculate one stud per running foot of wall to figure in blocking and built-up corners. To estimate roofing, which is sold by the square foot, find out how many square feet you need to cover, then add 10 percent.

When considering new or used materials, I recommend a balance between ease of construction and low cost. With this in mind, get new materials for the framing, sheathing, and roofing of your structure. You might find just the right used windows, but if you can't, try to find barn sash windows, which are inexpensive and scaled nicely, and they look good. Of course, higher-priced standard sash and casement windows of all sizes are available, but try to scale your windows to the size of your space.

CAD

If you are familiar with computer-aided design (CAD), a scaled floor plan might be easier for you to produce on a CAD program. I still use the simple (some say primitive) pencil-lines-on-paper method, which I believe is more efficient for a small-scale project. It's completely your choice!

·CHAPTER 9·

Building the Foundation, Sills, and Floor

You've found a suitable building site, checked with your city (and neighborhood owners' association, if necessary), and drawn up your plan. Besides the materials listed below, you'll need the tools listed in Chapter 1. You'll also need a good set of tall sawhorses. You can build your own sawhorses, as described in Chapter 1, but add 6 to 8 inches to the length of the legs. You'll also find a shovel and a wheelbarrow helpful for digging, hauling gravel, and mixing concrete (optional).

FOUNDATION AND FLOORING MATERIALS

PART	QUANTITY	DESCRIPTION
FOUNDATION BASE	12	⅔-cubic-foot bags ½-inch to ¾-inch crushed rock *or* gravel (8 cubic feet total); *or*
	12	similar-size bags dry premixed concrete
TEMPORARY LAYOUT STAKES	4	of any size
FOUNDATION BLOCKS	6	concrete pier blocks (see the drawing below)
FOUNDATION SILLS	2	4×6s, 16 feet long, pressure-treated
RIM JOISTS	2	2×6s, 16 feet long
FLOOR JOISTS	14	2×6s, 8 feet long
POSTS	1 or 2	4×4s, 8 feet long, pressure-treated (optional; you'll need these if the ground is sloped)
PERMANENT BRACING	2 or 3	1×6s, 8 feet long, pressure-treated (optional; you'll need these if the ground is sloped)
BATTER BOARD STAKES	12	cut from 2×4s, each stake from 2 to 4 feet long
BATTER BOARDS AND TEMPORARY BRACING	6	1×3 or 1×4 furring strips, 8 feet long
FLOORBOARDS	4	4 × 8-foot sheets ¾-inch-thick tongue-and-groove OSB or plywood
ADHESIVE	4	tubes construction adhesive
NAILS OR SCREWS	5 pounds	16d galvanized box nails for the joists; 8d coated sinkers or 2½-inch construction screws for the flooring
NAILS	2 pounds	6d coated sinkers
NAILS OR SCREWS	1 pound	6d or 8d duplex nails *or* 1⅝-inch construction screws for the batter boards

Common Types of Pier Block

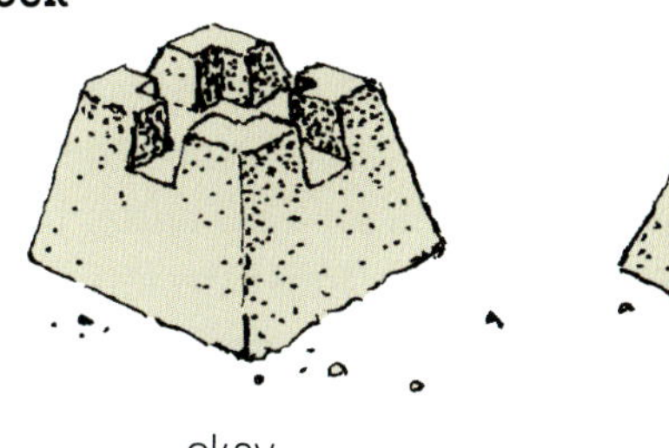
okay

better

okay

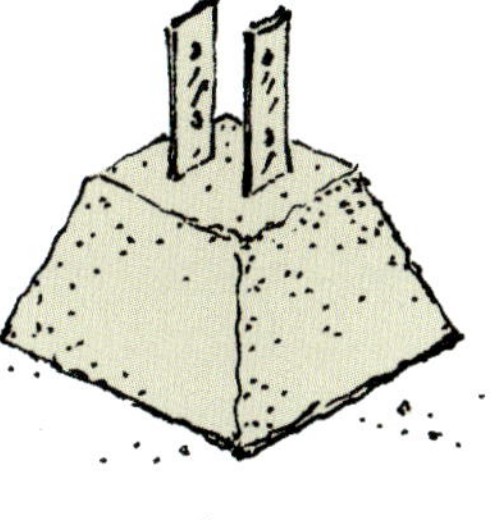
better

Building the Foundation

The foundation will consist of six concrete blocks set on gravel or concrete footings.

SET THE STAKES

Step 1. To lay out the foundation approximately, get your tape measure, four temporary stakes, and a hammer. Drive in a stake at each of the corners of an 8 × 15-foot rectangle.

Step 2. Have someone help you measure the diagonals (from corner to corner), adjusting the stakes as needed until the diagonal measurements are equal. If the ground is level, the diagonals should come out to 17 feet (204"). At this point, the measurement doesn't have to be exact; it can be off by an inch or so.

Step 3. Double-check to make sure the stakes are where you want the corners of your building to be.

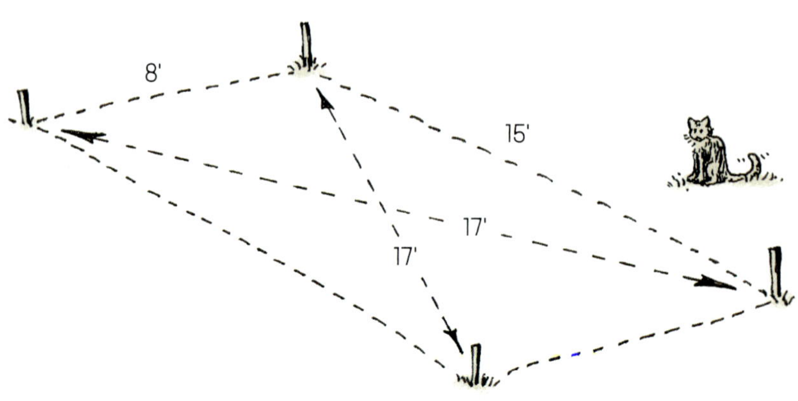

PUT UP BATTER BOARDS

Although these stakes give you an approximate location for your corners, you'll want to find the exact location by using batter boards and string. Batter boards are temporary L-shaped contraptions that help you find the exact corners of your foundation. The following steps may seem fussy, but if your foundation is crooked, your whole project will be thrown off.

Step 1. Make (or buy) twelve 2×4 stakes about 2 feet to 4 feet long, depending on whether your site is sloped. Cut eight batter boards, each about 3 feet long, from the 1×3 furring strips.

Step 2. Measure about 2 feet out from each temporary stake, and pound in three 2×4 stakes in an L formation that encloses the small stake. If your site appears to slope, drive shorter stakes at the uppermost corner and longer ones at the lower corners.

Step 3. Nail two batter boards to the 2×4 stakes in the highest corner of the foundation. Use 6d or 8d duplex nails, or 1⅝-inch-long construction screws. The batter boards should form an L shape and should be level.

Step 4. The batter boards at all the other corners must be level with the batter boards at this first corner. Attach a string to the top of the first batter board, then pull the string to the next corner, aligning it over the temporary stakes. Have a helper hook a line level to the middle of the string. As they watch the level, move the string up or down between the 2×4 stakes, keeping it aligned over the temporary stake. When the string is level, set a 1×3 batter board in place under the string. Have your helper mark the location of its edges on the 2×4 stakes. Then nail or screw it into place.

Step 5. Complete the L formation at this second corner by setting and nailing a second batter board level to the first.

Step 6. Repeat this procedure to install the batter boards in the remaining two corners of your foundation. Be patient and remember that this is only a "rough leveling"—it can be off by an inch or two and still work.

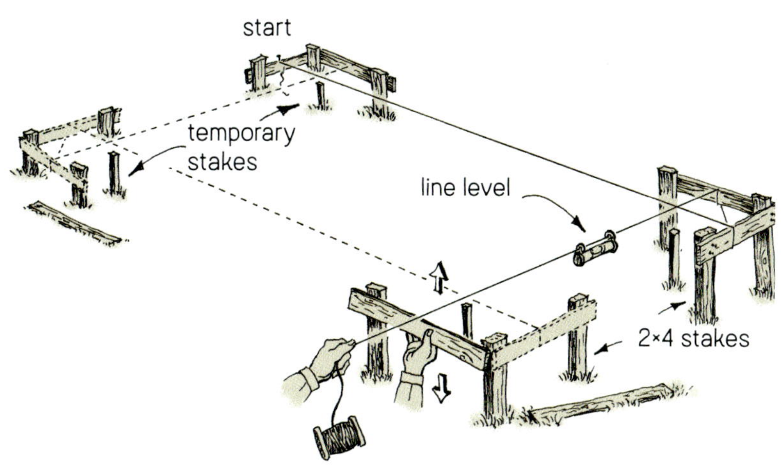

SQUARE AND MARK THE STRING

Now that the batter boards are up, you'll need to adjust the string and find the true corners of your foundation.

Step 1. Measure your foundation dimensions at the points where the strings cross in each corner. Have a helper hold the tape hook as steady as possible at one corner while you pull the tape to the diagonally opposed corner.

Step 2. Move the strings by sliding them along the batter boards, and check again with the tape, until the foundation rectangle is exactly 8 feet wide × 15 feet long and the diagonals in both directions are 17 feet. This can be a bit tricky, but don't pull your hair out—you'll get it eventually.

Step 3. Once the strings are set, clearly mark each point where the string meets the batter boards (just in case someone trips over the string later). A pencil mark will work, or use a nail to hold the string in place. With the corners squared and marked, now you can remove the temporary stakes.

Step 4. Mark the location of the two center piers at the midpoint of the long sides (7½ feet from each corner).

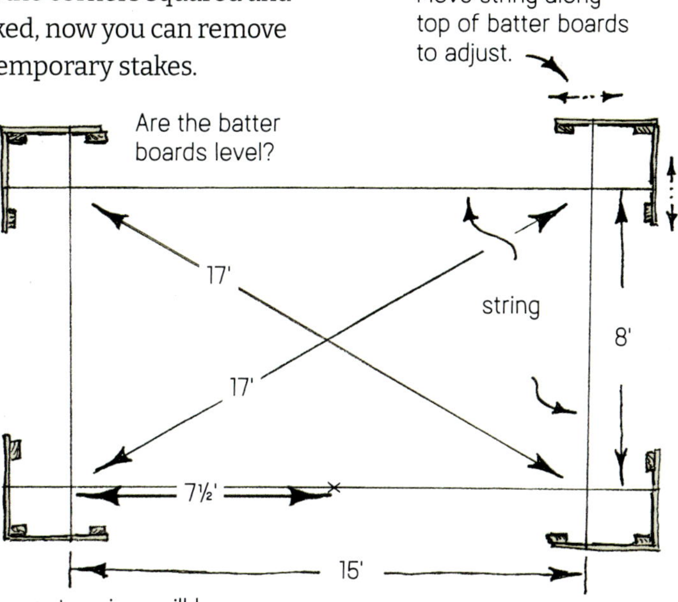

The center piers will be centered on the long sides.

SET THE PIER BLOCKS

You will set six pier blocks on gravel, or concrete if you wish (see the box on page 113), using your strings to get them in the right place. The point where the strings intersect is the exact corner of the building, which is also the outer corner of the post pocket in each corner pier block (see the drawing below). With this in mind, the footing holes are centered slightly inside the string corners. Here's how to set the blocks:

Step 1. At each corner, measure out a 16-inch square on the ground, centered about 2 inches inside the corner of the string. The two side holes are centered at the 7½-foot mark and about 2 inches inside the string.

Step 2. Using a spade, dig each square hole out to 8 inches deep, then fill it to the top with gravel (or concrete).

Step 3. As you set a pier block, have a helper lower a plumb bob from the string corner to the pier block while you jiggle the block into place, firmly setting it into the packed gravel, as shown. Level the pier block using your 9-inch level.

Step 4. For the side pier blocks, set the edge of the block's post pocket directly under the plumb bob and at the 7½-foot mark.

Step 5. Repeat for all the piers. Leave the strings in place.

Pier Block Layout

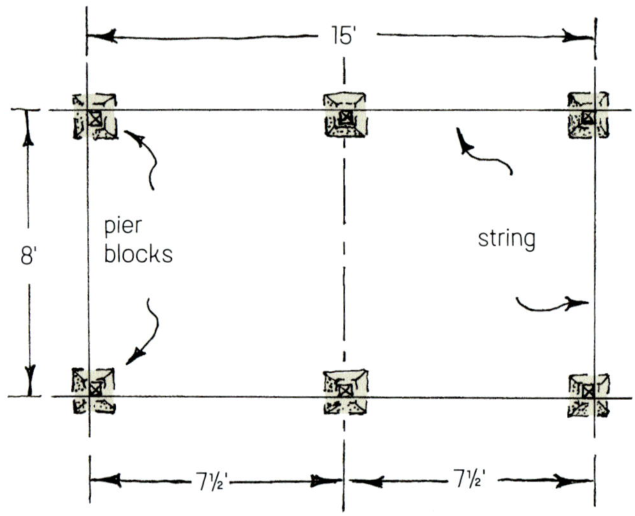

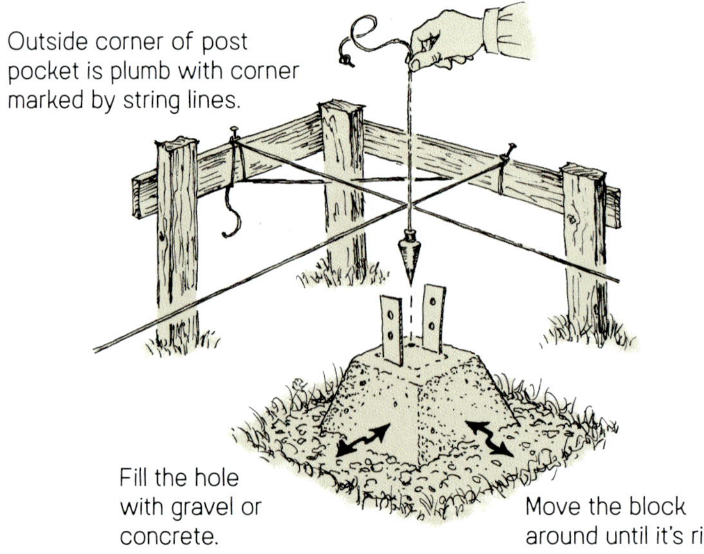

Concrete Footings for Uneven Ground, High Winds, Earthquakes, and the Like

For a sloped site, or if frost, loose soil, or earthquakes are an issue, or if your dwelling is exposed to high winds, you'll want to hold down the structure as securely as possible. Setting the pier blocks in concrete is one option. To set pier blocks, use one or two 50-pound bags of premixed concrete per pier block footing. Dump the concrete mix into a clean wheelbarrow, then add water according to the instructions on the bag. Mix one bag at a time at first, using a shovel or a garden hoe. Fill each hole with concrete, then immediately set the pier block as described in Step 3 on the facing page.

An alternative is to buy concrete tube forms to add depth and strength to the foundation. These are set in place and filled to the top with concrete. A steel bracket or post-tie is added to secure the post to the concrete. There are several brands available, which are sold in 4-foot lengths. A 7- or 8-inch-diameter tube will work for this project. You can easily saw them to any length. A Simpson Strong-Tie BC ZMAX or similar galvanized tie will work well for a 4-inch post.

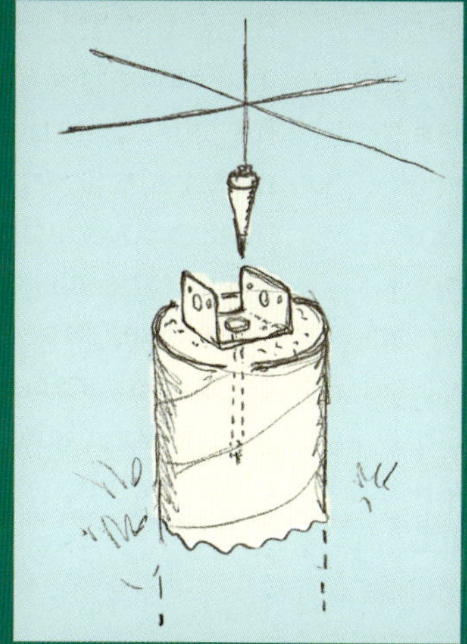

To set the tubes, dig down to the required depth (depending on your project), then position the tube under the foundation strings sticking 6 to 12 inches above the surface. Fill the tube with concrete. Using the strings as a guide, set the bracket or post-tie (see drawing). Insert a ⅜ × 6-inch-long bolt (not included with the tube) through the hole in the bracket into the concrete.

When working with concrete, always hose off the wheelbarrow and tools thoroughly when you are finished.

Making the Sills

Your floor will be supported by two 4×6 beams, or sills. If the ground is uneven or sloping, you may need to add posts to support the sills.

SET THE SILLS

Step 1. Measure and cut two pressure-treated 4×6 sills to exactly 15 feet long. This is a big cut, so draw your cut mark all the way around the sills. If you're hand sawing, watch the cut lines on the sides as well as the top for a square cut. With a power saw, cut across the top cut mark, then rotate the sill 90 degrees, saw on through, then rotate it and saw again. You'll get a clean cut this way.

Step 2. Set one end of the sill on the highest pier block and have a helper move it up and down at the other end while you eyeball your level. When it's level, measure the distance from the underside of the sill to the top of the lower pier blocks. If this distance is only a couple of inches, you can settle the highest pier block lower into the gravel until the sill is resting level on the next pier block. If the distance is 2 inches or more, you'll need to support the sill with posts at the lower pier blocks. Just cut a post to length out of a pressure-treated 4×4 (or from the 4×6 scraps you just cut off).

Step 3. Check for level at the middle pier block in the same way, and cut a post to length if necessary.

Step 4. To level the opposite sill, use an 8-foot-long straight board with a level strapped to it as a homemade long level. While your helper holds the level, measure the distance from the top of each pier block to the bottom edge of the level. Account for the height of the sill. If the remaining distance is greater than 2 inches, cut a post to size. If it's less than 2 inches, you can probably level the pier block simply by shifting it in the gravel base, as long as the gravel remains firmly packed under the pier block.

Step 5. With your helper's assistance, set up and secure the sills to the pier blocks as necessary. Toenail them to the posts with 8d galvanized box nails. If the sills want to lean or fall over, brace them with temporary stakes, as shown below. Don't take the strings down just yet.

Step 6. If any of the posts are more than 18 inches tall, brace them to your sills with permanent brace boards (see the drawing on page 116). For brace boards, use pressure-treated 1×4s or, better yet, 1×6s. Nail the braces in place with several 8d galvanized box nails at each end. As you nail them, check the sills to make sure they are in line with your strings. Leave any temporary braces in place for now.

Step 7. When your sills are braced, remove the strings and batter boards.

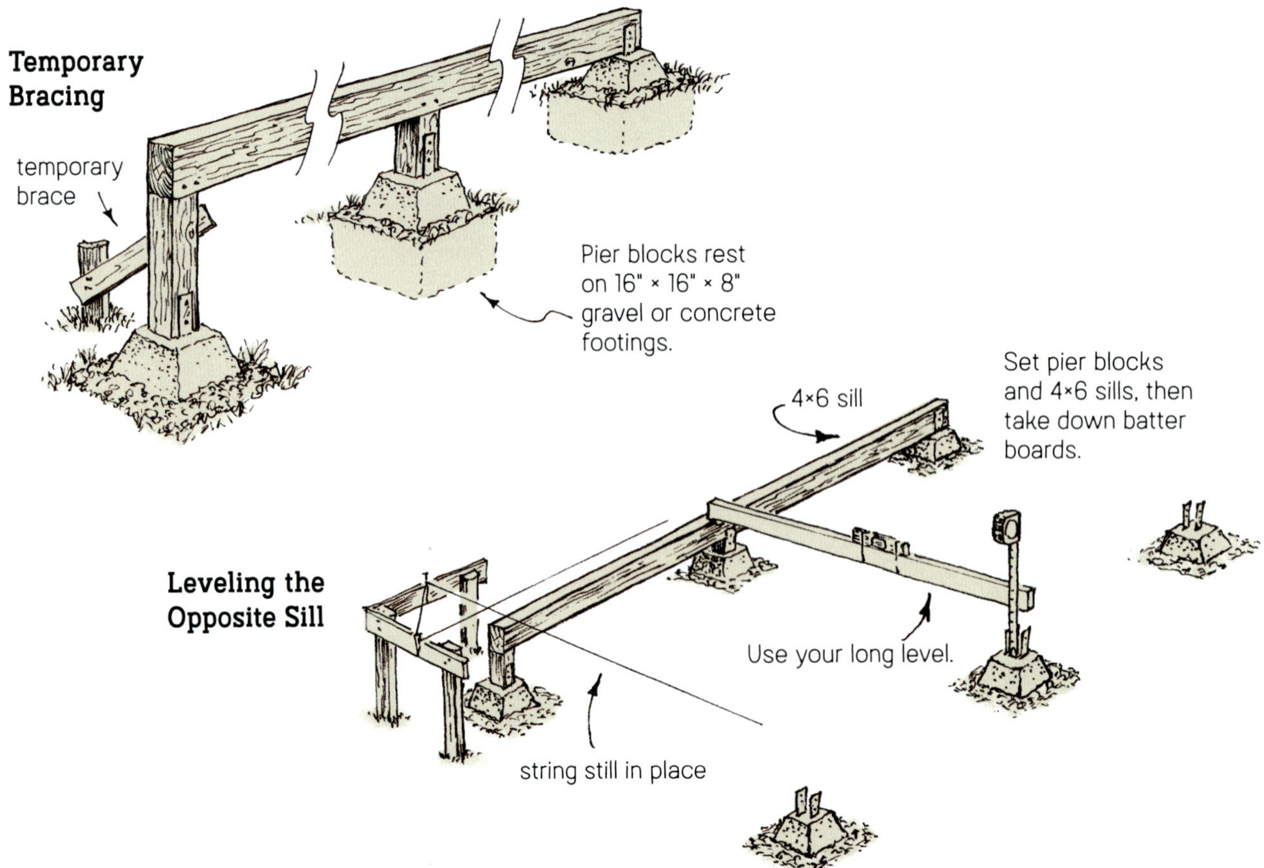

Constructing the Floor

Your backyard hideaway will have a floor made of 2×6 joists covered with sheets of tongue-and-groove OSB flooring.

INSTALL THE JOISTS

Step 1. Cut two long 2×6s to 15 feet (180") long; these will be the rim joists.

Step 2. Lay the two rim joists on edge, side by side on your sawhorses. With your tape measure, make centerline marks every 16 inches along the top edge of both boards.

Step 3. Measure to the left of each centerline and make edge marks on both boards. Using your square, draw edge lines (marking the edge of a joist) across both boards, on both the edge and the flat side. Mark an X next to each edge line, and over the centerline, to show which side of the edge marks the joists will meet (see the drawing at right). For added strength, you'll double the joists at the ends, so mark two Xs at each end of the joist.

Step 4. Set one rim joist on edge on the outer edge of a 4×6 sill (see drawing on page 116), and toenail it to the sill. The X's you drew on the board should face inside the foundation.

Step 5. Cut fourteen 2×6 joists to 7 feet 9 inches (93") long and place them on the sills at the X's. Sight along the edge of each joist with your eye. If it's bent or bowed, turn it so it bends up in the middle. The joists should end 1½ inches from the outer edge of the other sill.

Step 6. Have a helper hold down each floor joist while you nail the rim joist into it with three 16d nails. A framing hammer is useful here.

Step 7. Set up the opposite rim joist and check to see that your marks line up the same way.

Step 8. Nail the rim joist to the ends of all the floor joists and toenail it into the sill. The joists should fit on the sills as shown on page 116.

Step 9. Add more permanent braces from the joists to the posts, if necessary, then remove any temporary braces.

Marking Floor Joist Locations on the Rim Joists

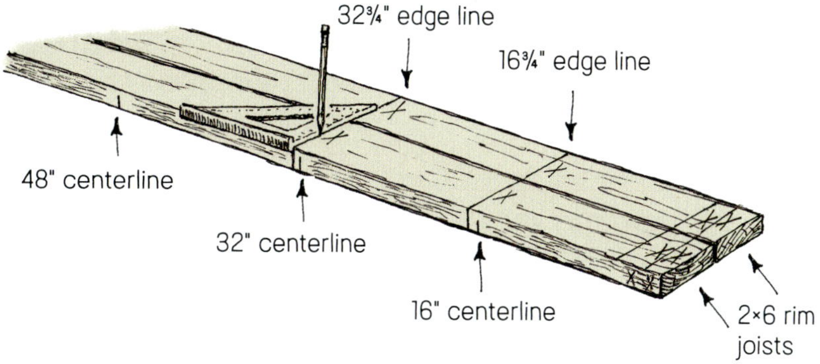

Assembling the Floor Frame

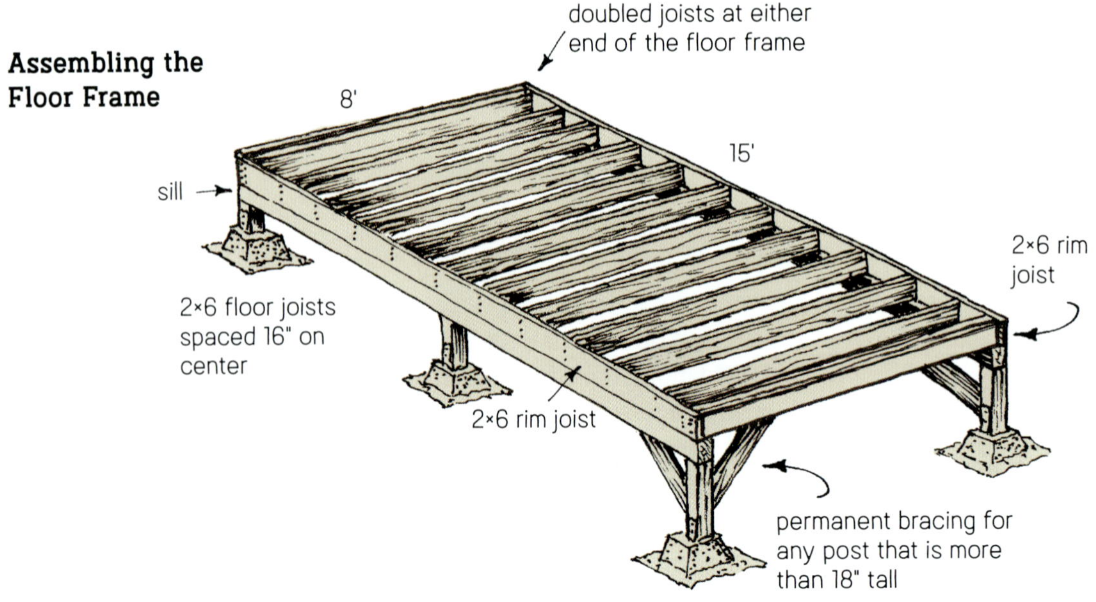

NAIL OR SCREW DOWN THE FLOOR

Now that all the joists are nailed together, the next step is to nail down the floor itself. I recommend ⅝- or ¾-inch-thick tongue-and-groove OSB in 4 × 8-foot sheets. It's both inexpensive and strong—a winning combination! Stagger the sheets for extra strength (see the facing page).

Step 1. Lay out a full sheet of OSB on your joists, line it up neatly to a corner, then lay another to the end of the floor. The two sheets should join over a joist. Using 8d nails, tack the sheets down so they won't be pushed out of place when you begin laying the next row of flooring.

Step 2. You'll want the seams in the flooring to be staggered, so begin the next row of flooring with a 48-inch-long sheet. Cut the sheet to size on sawhorses or on the floor. Set your circular saw blade to cut 1 inch deep or ¼ inch below your floor panel. Then use a scrap board under your cut to avoid sawing into the floor.

Step 3. Continue laying and tacking down OSB, staggering the seams, until your floor is covered. If the tongue-and-groove edges don't join easily, use a 2×4 block and a hammer to nudge the sheets together.

Step 4. Saw off any overhanging scraps while the sheets are in place, so that the panels are now all tacked down and lined up evenly with the outer edge of the floor framing. This is where foundation fussiness begins to pay off!

Step 5. Nail or screw down the flooring with 8d coated sinkers, galvanized box nails, or 2½-inch-long construction screws. Use your regular (not framing) hammer for this. Use one nail every 6 inches on the edges and one every 12 inches on every inside joist. (To guide your nailing into the inside joists, you can mark the location of the joists right on the panels.)

Nailing Down the Floor Panels

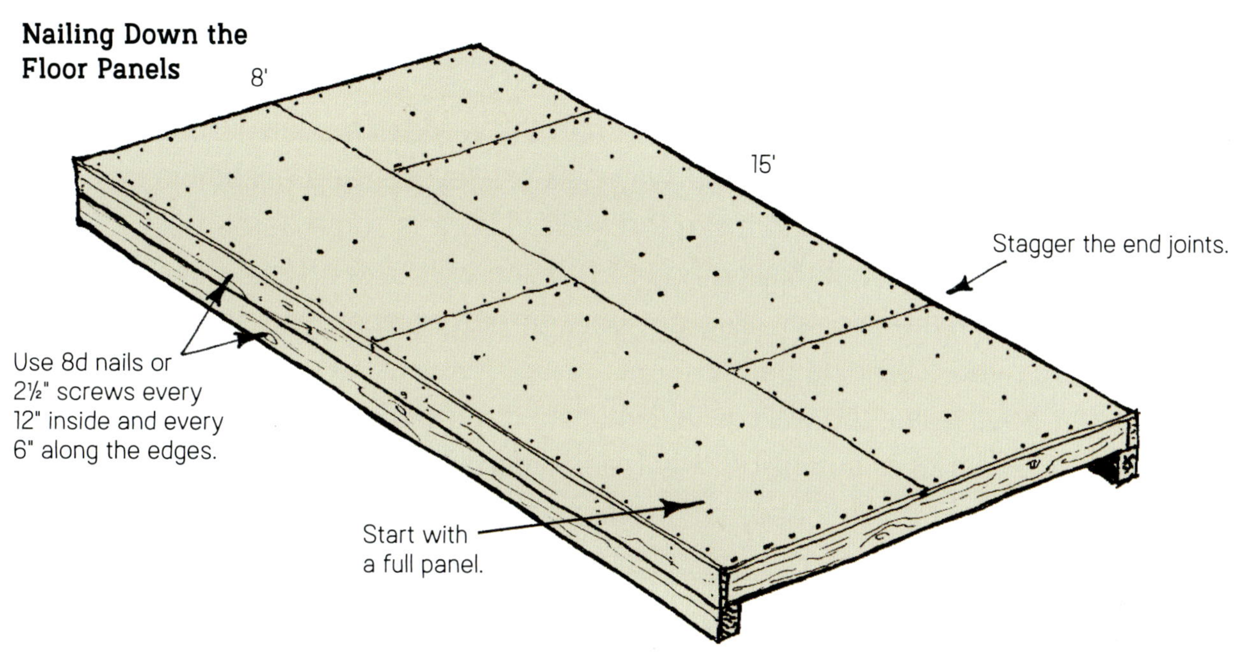

Heavier Hammering

For framing, I recommend a 22-ounce framing hammer with a smooth (not waffle) head. Once you've set a 16d nail, hold the hammer handle well away from the head and swing with your whole arm to drive in the nail. With fewer hammer strokes, the nail will have less of a chance to bend (and you will last longer as well!). It may take some practice to get the feel of your hammer.

·CHAPTER 10·

Building the Walls

With the framing of the walls, we enter the world of modern house carpentry. I'll show you the way builders lay out and assemble the pieces of wood to make the walls. There is some carpenter language here, but this method of framing will make efficient use of your materials and also get your structure approved should you need a building permit.

You'll frame the walls with 2×4 studs centered every 16 inches. To save a lot of cutting of 2×4s, I recommend buying the 92⅝-inch-long precut studs. These will make a framed wall a little more than 8 feet tall—the standard wall height in house construction. Set aside leftover 2×4s and sheathing for the gable-end walls, which you will construct after the roof is on.

WALL MATERIALS

PART	QUANTITY	DESCRIPTION
PLATES	10	2×4s, 8 feet long
PLATES	6	2×4s, 16 feet long
WALL STUDS	60	2×4s, precut to 92⅝ inches long
HEADERS	3	4×4s, 8 feet long
BRACING AND TRIM BOARDS	8	1×4s, 12 feet long
SHEATHING	14	4 × 8-foot sheets 7⁄16-inch- or ½-inch-thick CDX plywood *or* OSB
NAILS	5 pounds of each	16d coated sinkers; 6d, 7d, *or* 8d coated sinkers

Cutting and Marking the Top and Bottom Plates

To begin building walls, you'll cut and mark the top and bottom plates. From the plan you drew, transfer the locations and dimensions of your door and windows directly to the plates. The marks will then tell you exactly where to nail on the studs (at the X's), the trimmers (at the T's), and the cripples (at the C's).

CUT THE PLATES

Step 1. Set up your sawhorses, and place four 16-foot-long 2×4s on them. Cut all four to 14 feet 5 inches (173") long. These will be the top and bottom plates for your longer side walls.

Step 2. Set aside four 8-foot-long 2×4s. These will be the top and bottom plates for your shorter end walls.

Step 3. Mark each pair of plates with a big 1, 2, 3, or 4 for each wall number you will build. Set aside the extra plate of each pair. Number the walls on your plan in the same order.

Step 4. Set one plate on the floor for each wall, as shown in the drawing below. Tack each plate down with a duplex nail or two if it won't stay put.

Wall-Plate Layout

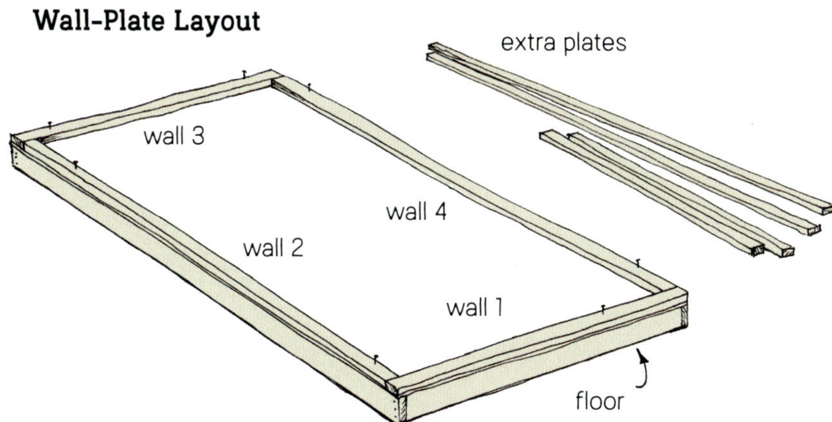

MARK THE PLATES

Now let's mark the plates where there will be a door, windows, or studs.

Step 1. Spread out your floor plan, with dimensions, on your new floor. Using your plan, measure from the corner of the building to the centers of your window and door locations and mark them on the plates. Some builders write CL on these marks to distinguish them from stud centerlines.

Step 2. From the window and door centerlines, measure to the edges of each door and window opening. Look at the plan to see how wide the opening is, then measure half that width to the edges. With your square, draw the edge lines on the plate. Next to each edge line and away from the center of the opening, write a T where a trimmer will go.

Step 3. Measure 1½ inches farther from the center of the opening, draw another edge line, and write an X on the other side of that edge line to mark where a stud will go. So now you have a line, a T, a line, and then an X, as shown in the drawing below.

Step 4. Measure the distance between your trimmer edge lines to check if that is the actual width of your window opening.

Step 5. For the studs, hook your tape measure to the corner of the building, and mark a centerline every 16 inches on the plate. Continue to the far end.

Step 6. Make edge marks ¾ inch past the centerlines, as you did for the floor joists, then draw X's over the centerline marks. Where a stud falls within a window or door opening, draw a C for cripple, which is a short 2×4 placed above or below a window. If a window edge is close to a stud, nudge the window opening over to meet the stud to save some lumber, *if* moving the window won't mess up your plan. Check your plan, then redraw your window edge marks carefully. Don't move the studs or cripples, or you'll have problems to contend with later.

Step 7. At each end of the plate, draw an X. Where two plates come together at a corner, draw an X at the end of one of them and three edge lines and X's on the other one, for what builders call a built-up corner (see the drawing below).

Step 8. Move on and mark all the plates. It will make more sense (and get easier) as you proceed.

Step 9. Once all the plates are marked and the window and door openings are the right size, get the second plates that you had set aside. Lay them flat and edge to edge with the plates you just marked. With your pencil and square, transfer all the marks to these blank plates. Clear all the plates off the floor except for those for wall 1. Now you are ready to cut the pieces and assemble the wall frames.

Marking the Wall Plates

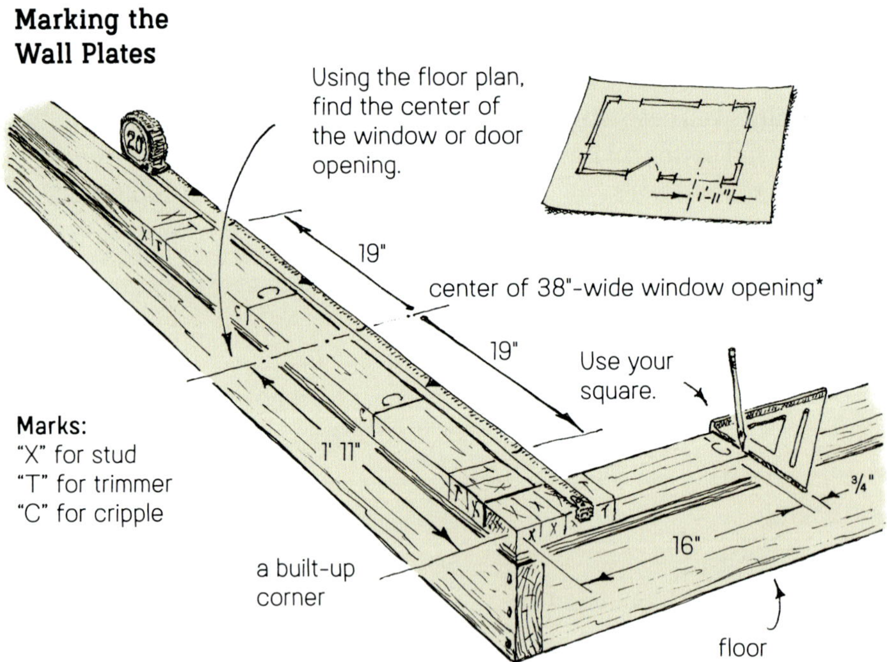

*The rough opening for a window is usually 2" wider than the actual window.

Building Wall 1

If you are still not sure about all those marks you just drew on the plates, keep going and things will become clearer. Beginning with wall 1, you'll assemble each wall on the floor, using your plan and the marks on the plates. The large window opening in wall 1 will be 56 inches wide and 38 inches tall, and the bottom of it will be 44 inches from the floor. With this information, you can figure out the lengths of all the pieces and then assemble the wall, starting with the studs.

ASSEMBLE THE STUDS AND PLATES

Step 1. On your floor, arrange the top and bottom plates on edge and about 8 feet apart for wall 1. The plate nearest the edge of the floor where this wall will go will be the *bottom* plate.

Step 2. Lay down a 92⅝-inch precut 2×4 stud over each X so that the layout begins to look like the drawing on page 123.

Step 3. The ends of this wall are built up, so you will need to assemble three 2×4 blocks (scraps are okay) inside of two studs, as shown at right. Use your square to check that the ends are square, and then drive three 16d nails through the sides of the studs into each block from both sides. Set the built-up ends in place between the plates.

Step 4. To start nailing the wall frame together, set a stud with its edge to the edge line and covering the X, and hold it in place with your foot or knee. With your framing hammer, drive two 16d nails through the plate into the stud. For the built-up ends, nail two nails into each stud and into the block.

Step 5. Continue until all the studs are nailed to the plate.

Step 6. Move to the other plate, and nail it to all the studs as well.

Nailing Tips

Wall frames are usually nailed on the floor, with pieces laid on edge. To start, step on the stud and set the nail into the plate. Then, with your foot on the stud, nail through the plate and into the stud, swinging with your whole arm for momentum (and to avoid fatigue). When the nail is through the plate, check to see that the pieces are flush and the stud/cripple/trimmer is lined up where it should be. Finish nailing it in. This takes a little practice, so be patient with yourself!

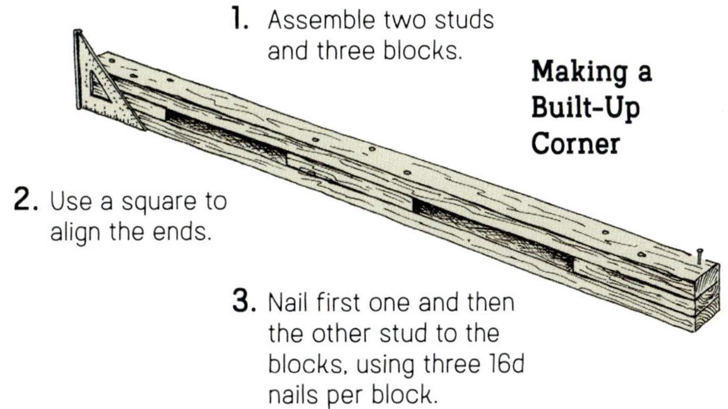

Making a Built-Up Corner

FRAME THE WINDOW

Next, you'll add the pieces that frame the window. These are the trimmers, window rough sills, window headers, and cripples.

Step 1. Go to the studs set next to the T marks on the plates. From the bottom of these studs, measure to 41 inches and then 82 inches, and at each location draw edge lines with your pencil and square. *Above* the 41-inch edge line, write an S for sill; *above* the 82-inch edge line, write an H for header. You just marked where the rough sill and the header will go for your window.

Step 2. Measure the distance along the bottom plate under the window to get the length of the window rough sill and header. The header should be 59 inches long, which is the width of the window plus 3 inches for the thickness of the two trimmers.

Step 3. Cut five 2×4s to 41 inches long for trimmers and cripples. Set all four into place on the T and C marks on the bottom plate. Nail the trimmers to the studs with 12d nails, then drive two nails through the bottom plate into each trimmer and cripple end.

Step 4. Cut a 2×4 to 59 inches long for a window rough sill. Set the sill in place on the trimmers and cripples. It should line up with the edge line you drew on the studs. Drive two nails through the studs into the sill ends and then two nails through the sill into each trimmer end and cripple end. Use 12d or 16d nails.

Step 5. Cut two 2×4s to 38 inches long for the upper trimmers. Set these on the sills and nail them into the studs with two or three 12d nails.

Step 6. Cut a 4×4 or two 2×4s to 59 inches long for your window header. With the 2×4s, make a header sandwich by nailing them together with filler scraps of ½-inch plywood or OSB to make the header 3½ inches thick.

Step 7. Set the header, with the 2×4 edges on the trimmers. Drive two or three nails through the studs into each end of the header.

Step 8. Cut five short trimmers and cripples that go between the top of the window header and the top plate. Measure the vertical distance, then cut four 2×4 blocks to fit. Drive two 12d nails through the top plate into each block. Toenail the blocks into the top of the header with 6d nails.

Step 9. Check your window opening once more. It should be 38 inches high by 56 inches long.

RAISE WALL 1

Step 1. To raise wall 1, first lift or pry the top plate off the floor enough to slide a 2×4 scrap under it. Then you can grab hold.

Step 2. Have a helper help lift the wall and nudge it to the edge of the floor. Plumb and brace the wall in place.

Step 3. Nail the bottom plate into the floor between each stud, using 16d nails.

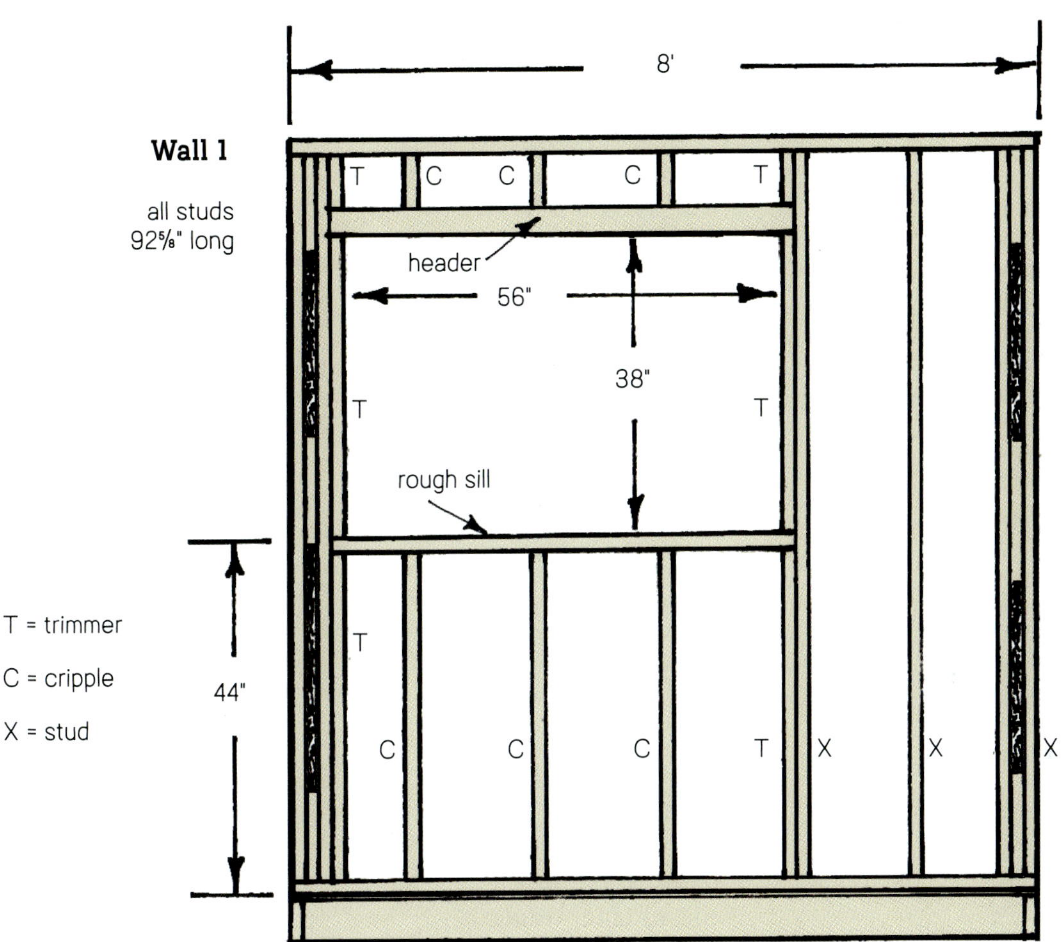
Wall 1
all studs
92⅝" long
T = trimmer
C = cripple
X = stud
8'
T
C
C
C
T
header
56"
38"
T
T
rough sill
44"
T
C
C
C
T
X
X
X

Building Wall 2

Frame the door on wall 2 with the following in mind. Assuming you have a prehung door that is 30 inches wide by 80 inches tall, you'll need to add 2½ inches to the height of the door opening (to accommodate the threshold) and 2 inches to the width, for the doorjamb and some wiggle room. Therefore, your door rough opening needs to be 82½ inches high and 32 inches wide. Since the bottom plate is 1½ inches thick, the trimmers will be 81 inches long. The header will be 35 inches long to include the thickness of the two trimmers.

RAISE WALL 2

The longer wall frames are heavy, so once you have nailed wall 2 together, have an assistant or two help you raise it.

Step 1. Lay the plates for wall 2 on the floor. Then cut the pieces and assemble wall 2 in the same way you did for wall 1.

Step 2. Lift or pry the top plate off the floor enough to slide a 2×4 scrap under it. With one or two people helping, lift the wall, nudge it to the edge of the floor, and set it snugly against wall 1.

Step 3. Tack a single 16d nail through the end stud of wall 2 into wall 1, then plumb and brace wall 2 as you did wall 1.

Step 4. Nail the bottom plate into the floor between each stud *except* in the door opening. Use 16d nails.

Step 5. Nail through the end stud of wall 2 firmly into the studs and blocks of the built-up corner of wall 1, using five or six 16d nails.

Step 6. Cut out the bottom plate between the doorway trimmers.

Note: Depending on your needs, you may want a 32- or 36-inch door.

Wall 2

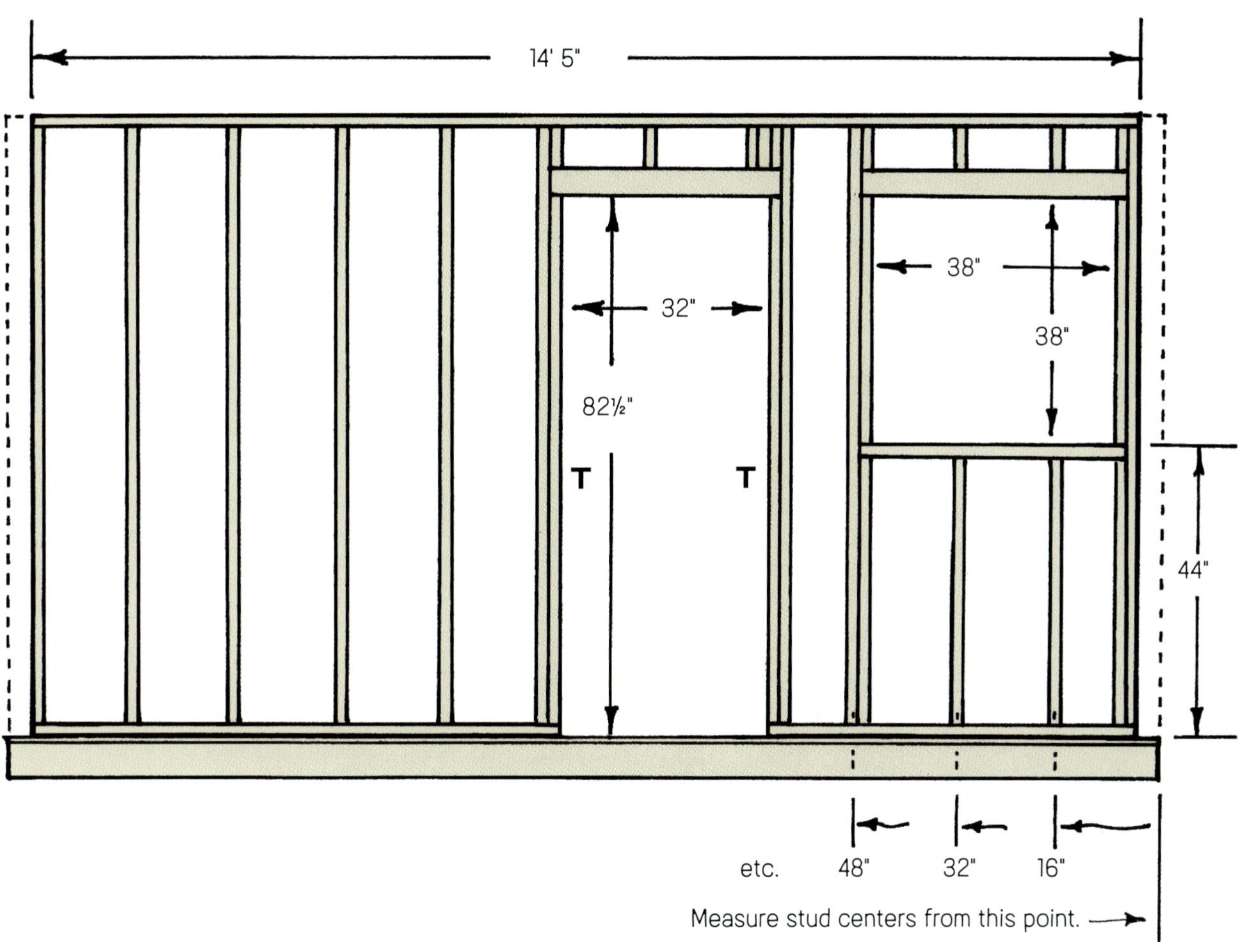

Choosing a Door

Most new doors are sold as prehung units, meaning they are already hung on hinges in a wooden frame or jamb. At the lumberyard, they will call a 30-inch-wide door a 2/6 door, meaning it is 2 feet 6 inches wide. If you find a really cool unframed door, you can make your own jamb for it and apply the hinges. This is discussed more on page 151. If this is your first door installation, I recommend buying a new prehung unit, which usually comes with its own installation instructions.

Building Wall 3

This wall has two small windows, a low one for the toilet room or closet and a high one that extends above the top plate for the loft. Look at your floor plan to find the window-opening centers, and look at your side views for the window-opening heights. Cut the pieces and assemble the wall frame, as you did with walls 1 and 2.

When you have marked both plates, take the top plate to your sawhorses and cut out the piece to allow for the frame for the loft window. The 35½-inch edge lines will be your cut marks. Then bring the pieces back to the floor to assemble the wall.

When wall 3 is framed, lift it up, tack it to wall 2, plumb and brace it, and nail it to the floor. Finally, with five or six 16d nails, attach it to wall 2.

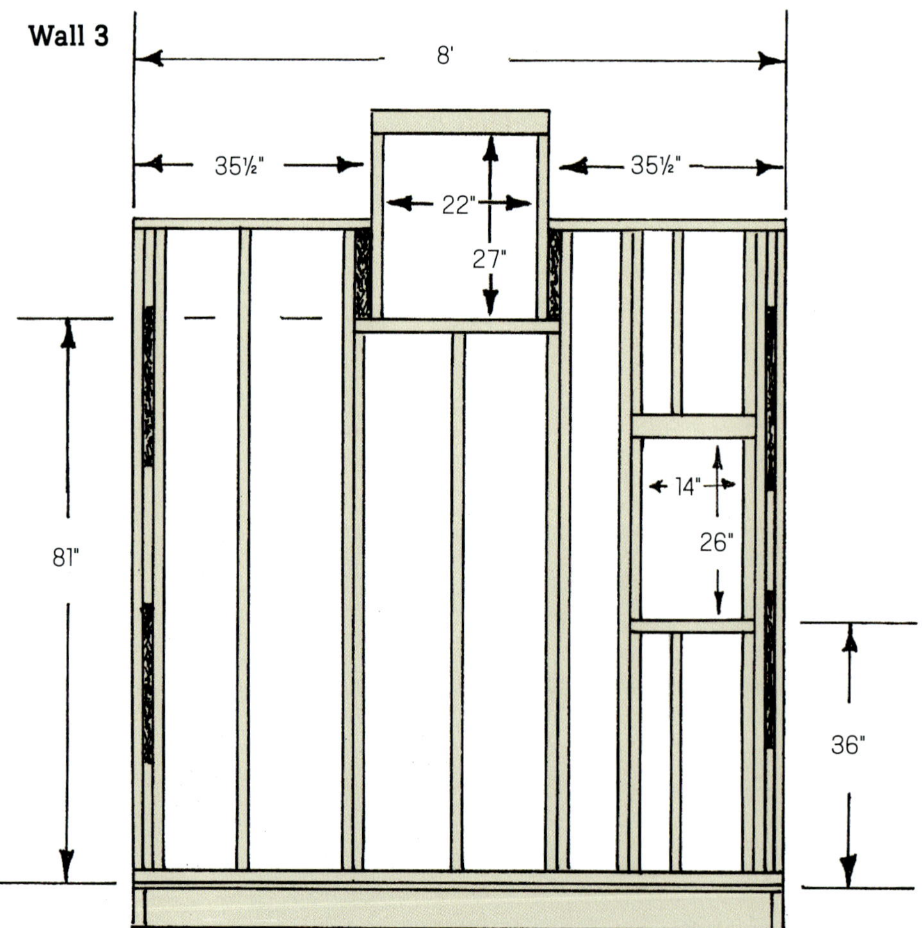

Building Wall 4

This wall has a low, under-loft window as well as a higher window for the kitchen counter. Depending on the size of the windows you've chosen, mark and cut the pieces and assemble the wall frame as shown in the wall 4 drawing below. Then lift it into place, tack it to both wall 3 and wall 1, brace it, and nail it first into the floor and then into the other walls.

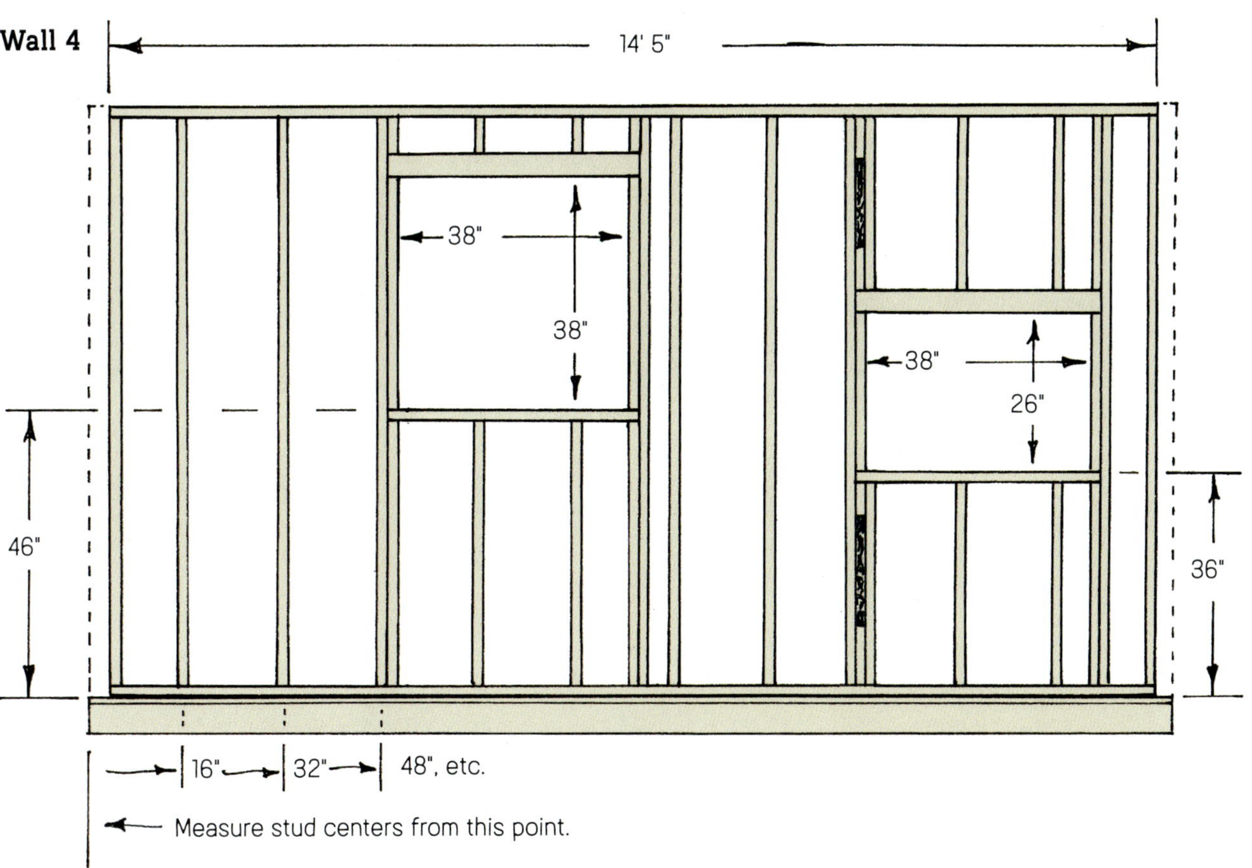

Plumb the Walls

As you build the walls, they may get knocked out of plumb by all the pounding and nudging. If so, they will need to be straightened out and then re-braced before you cover the walls with sheathing.

PLUMB AND RE-BRACE THE WALLS

Step 1. First, check that the tops of all the plates are even (flush with each other) at the corners. If they're not, pound down any high places to even them out.

Step 2. Check the wall corners to see if they are still plumb. If they're not, loosen up the braces so you can re-plumb them.

Step 3. Have a helper hold one of the loosened brace boards, ready to nail it in. Have another helper gently nudge the wall back to plumb, while you hold your long level to an outside corner. When it looks good on both sides, tell them to hold it while you check for plumb. If it still looks good, nail it in.

Step 4. Repeat until all the corners are plumb in both directions.

Step 5. If the walls bow in or out at the top, use a cross brace from the top of a stud to the bottom of a stud on the opposing wall, or brace it to a stake in the ground.

Bracing the Walls

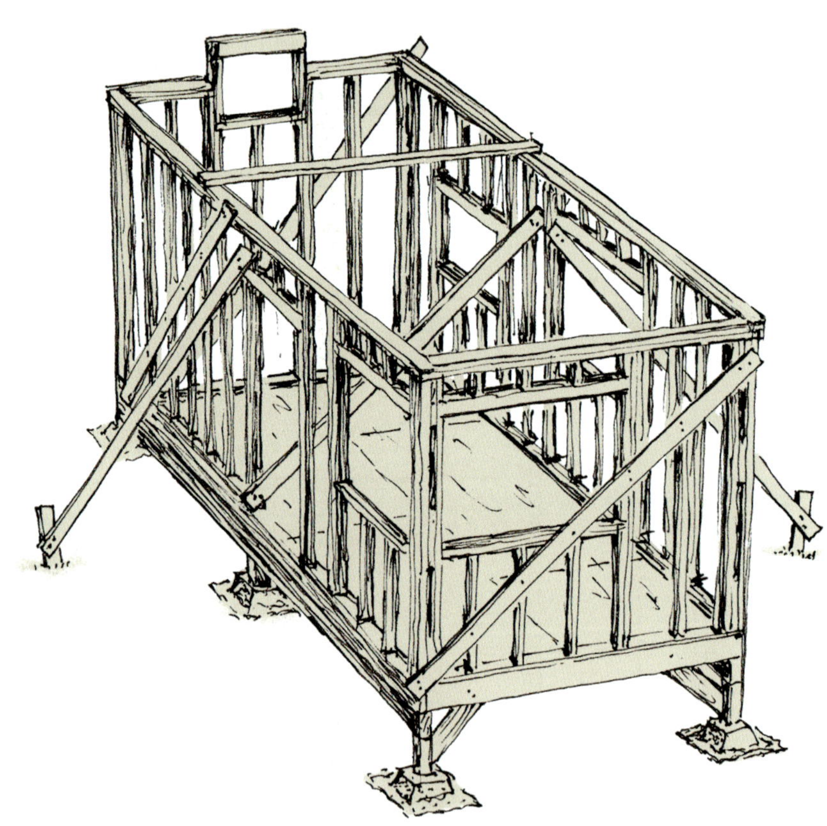

Installing the Second Top Plates

This second layer of top plates will tie all the walls together and help straighten them out. We will cut the top-plate boards so they overlap the original top plates at the corners.

ATTACH THE SECOND PLATES

Step 1. Cut two 2×4 pieces 15 feet (180") long and two pieces 7 feet 5 inches (89") long.

Step 2. Set the plates flat on top of your walls. (You might have to loosen a brace or two.) If a wall should happen to bend or bow in, and the top plate also bends a bit, set the top plate so that it bends the opposite way to cancel out the bow in the wall.

Step 3. Nail the second top plates onto the top of the original plates with two 12d or 16d nails every 16 to 24 inches. Use three nails at the overlapped corners to tie the walls adequately. If necessary, add a brace board to the floor or across the plates to keep the long walls parallel.

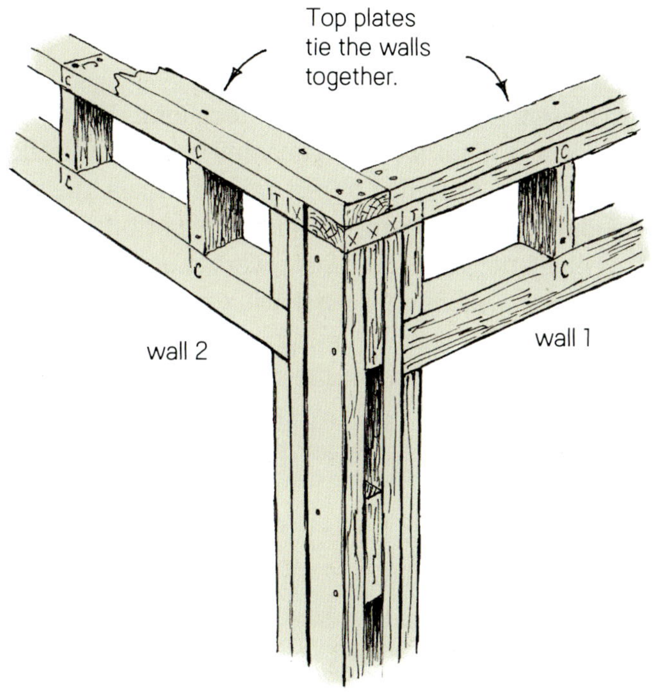

Covering the Walls

I recommend covering the outside of the walls with ½-inch-thick CDX plywood or OSB for economy and strength. Either material is strong enough to withstand earthquakes, high winds, or a trip down the highway, should you ever want to move your shelter. OSB is the less expensive material, but plywood accepts shingles or siding a little better. Later, you'll add the exterior siding of your choice (see Chapter 7) and interior pine paneling or drywall.

If you laid out your studs on 16-inch centers, you should be able to nail up your panels edge to edge without any trimming.

SETTING UP A PANEL REST

There's a trick to getting the most out of 8-foot-long sheathing panels: Get a long, straight 1×4 or 2×4 and nail it to the outer floor joist or rim joist to make a ledge, or rest, for the panels. The top of the rest should be about 5¼ inches above the bottom of the floor joist, or high enough so the panel covers at least the lower half of the top plate. This way, the panel will reach both the lower and upper plates to help brace your framing (see the drawing on page 131). Later, after you install the panels, you can cover the remainder of the joist with sheathing scraps. The roof and a soffit board will cover any gap left at the very top.

MARK AND CUT THE PANELS

Step 1. At the end of a wall, set a panel on the panel rest and have a helper hold the panel to the wall, or tack it to the wall with a couple of nails.

Step 2. Now go inside and use a pencil to trace the edges of any window or door openings onto the panel. These will be your cut lines. Also check to make sure the panel's long vertical edge is centered on a stud so that the next panel will have a nailing surface.

Step 3. Take the panel down and place it on your sawhorses. Cut out the window opening with your circular saw. You may want to set two wide planks on your sawhorses to better support the panel. To avoid deep cuts in the planks, adjust your saw blade so that it cuts only ⅛ inch deeper than the thickness of the panel.

Working with 9-Foot Panels

Plywood and OSB siding are also available in 9-foot-long panels. If you use these, cut them to 8 feet 8 inches long. Set your panel rest to ½ inch *below* the bottom of your floor joists and nail the panels in place. The panels should then reach the top of your wall frame. If this is your finished siding and it's exposed to the weather, use galvanized 8d box nails or special siding nails.

NAIL IN THE PANELS

Step 1. Set the panel back on the rest and check to see that it fits. For a superstrong bond, apply adhesive (such as Liquid Nails) on the studs and at the very top and bottom of the panel before nailing.

Step 2. Nail the panel in place with 6d, 7d, or 8d coated sinkers. Drive a nail every 6 inches along all edges of the panel and every 12 inches over the center studs. (To find the center studs, measure across the panel to 16 and 32 inches.) Try to angle some nails into the floor joist, if possible, since there is not much nailing surface there. Use your lighter-weight hammer for these nails.

Step 3. Once you have finished paneling one wall, remove the braces from the next wall, panel it, and so on. If the rim joist sides are exposed, remember to cover them with scrap pieces of paneling so your building is evenly covered. Cover the joists only to their bottom edge.

Step 4. On the gable-end walls, leave off any paneling above the windows. You can put up these pieces once you complete the upper framing for the gable ends.

Putting Up Wall Sheathing

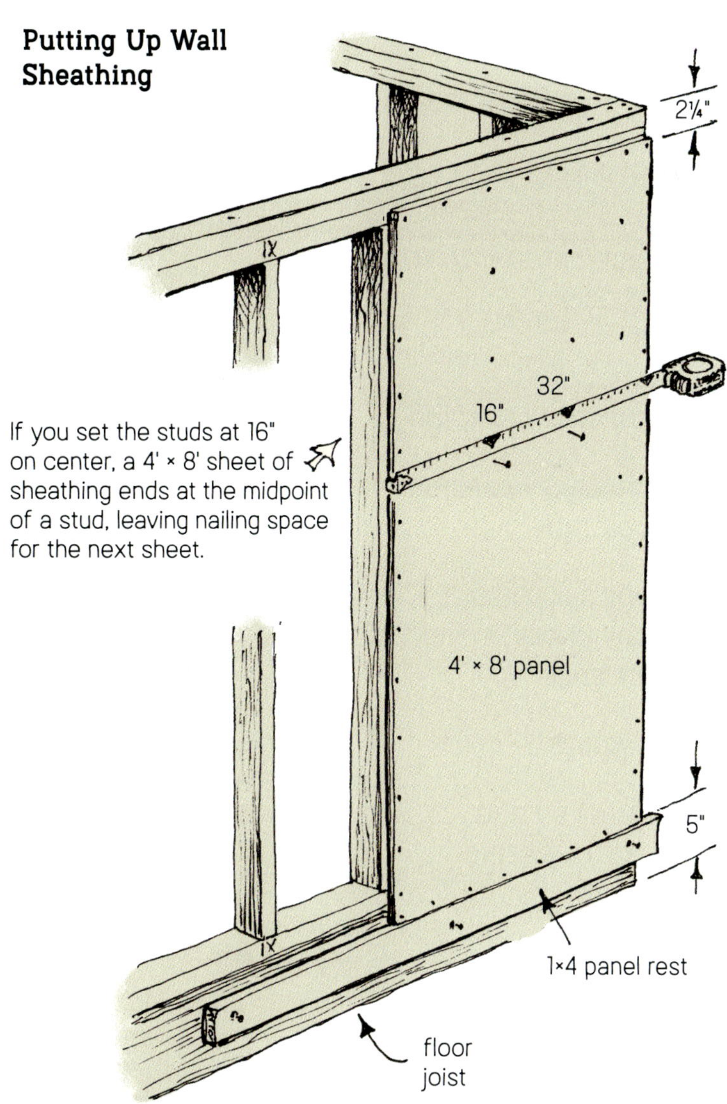

·CHAPTER 11·

Building the Roof

Now let's take a look at the roof. The Classic Design in Part 1 uses a basic boards-on-beams style of roof, which is fine but limited to a very small structure. The backyard dwelling you're building here has a gable roof that is framed cathedral-style to give the inside a spacious feeling and provide room for a loft.

ROOF MATERIALS

PART	QUANTITY	DESCRIPTION
RIDGE AND FASCIA BOARDS	3	2×6s, 18 feet long
RAFTERS	14	2×6s, 12 feet long; *or*
	28	2×6s, 6 feet long
COLLAR TIES	2	2×6s, 8 feet long
TEMPORARY BRACING	6	1×4s, 8 to 12 feet long (you can use the same temporary braces you used on the walls)
TEMPORARY BRACING	4	2×4s, 12 feet long (you can use these later for the loft)
ROOF TRIM	2	1×8 pine or cedar boards, 12 feet long (optional; see Trimming the Roof on page 143)
ROOF SHEATHING	7	4 × 8-foot sheets ½-inch-thick plywood or OSB
ROOF UNDERLAYMENT	1 roll	15-pound felt
DRIP EDGE	6	10-foot lengths 1½-inch × 2⅝-inch aluminum drip edge (see the drawing on page 141)
SHINGLES	2 squares (200 sq. ft.)	three-tab or similar asphalt shingles
RIDGE VENTING	16 linear feet	shingle-over style
SOFFIT VENTING	1 roll	3-feet-wide × 5-feet-long × ¼-inch- or ⅛-inch-mesh galvanized wire hardware cloth
SOFFITS	4	1×8 pine or cedar boards, 10 feet long
NAILS	2 pounds	16d galvanized box nails
NAILS	1 pound of each	6d galvanized box nails, for the trim boards; 6d duplex nails, for the brace boards; ¾-inch or 1-inch galvanized roofing nails
NAILS	5 pounds	6d, 7d, or 8d galvanized box nails

Marking and Cutting the Roof Framing

This gable roof will have a 6:12 pitch, which will shed water nicely but won't be too steep to work on. For convenience, you can reduce this proportion to 1:2. The roof has a span of 8 feet, so half of that (the run of a rafter) is 4 feet. From your 1:2 proportion, you can find the rise, which will be 2 feet.

With this valuable information, you can now determine the length of your rafter. One way is to use the Pythagorean theorem; however, that would require adjustments. Another method is to draw a rafter diagram to scale using drafting tools and an architect's scale. I recommend the 1-inch scale. This way, you can draw out your rafter to its exact shape, then measure its length with your architect's scale.

The diagram method will also help you determine the details of your rafters. As you can see in the drawing below, the rafter meets the ridge board, and the thickness of the ridge board will shorten the rafter length. Conversely, the part of the rafter that extends beyond the wall, known as the tail, will lengthen the rafter, depending on how much overhang you want.

About Roof Pitch

A gable roof is built at a pitch, or angle. Pitch is a proportion that tells you how steep a roof is. A standard roof pitch varies from 1:12 to 12:12, and sometimes steeper. The first number is the height of the rise, or the vertical dimension; the second number is the length of the run, or the horizontal dimension. If the pitch is 1:12, then the roof rises 1 foot over a 12-foot run and is nearly flat. If the pitch is 12:12, then the roof rises 12 feet over a 12-foot run and is quite steep, which is appropriate for snowy winters.

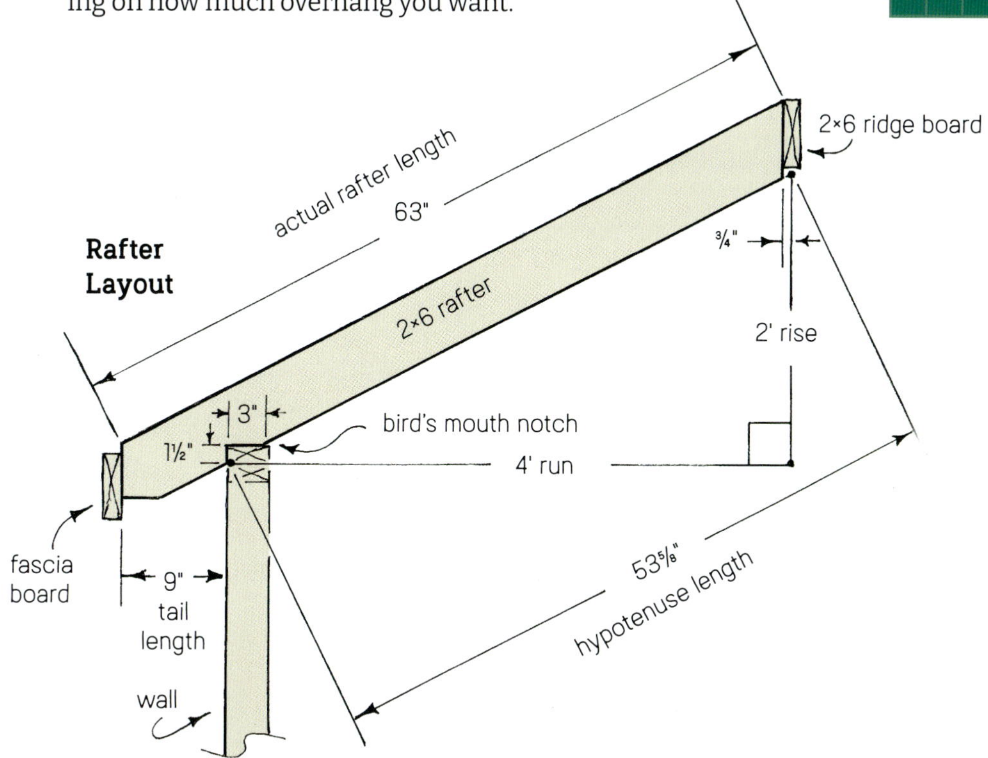

You'll cut a notch called a "bird's mouth" into each rafter so that it will rest solidly on the top plate of the wall. If sheathing covers the wall framing, remember to add its thickness to the width of the bird's mouth.

Now you can measure and cut out a rafter. In this demonstration, the sheathing is *not* covering the wall at the bird's mouth. If there is sheathing, add its thickness to your bird's-mouth cutout.

MARK AND CUT THE TEST RAFTERS

Step 1. To start out, set your rafter square at one end of a 6-foot length of 2×6.

Step 2. Looking at the rows of numbers, or scales, on the square, find the "common" scale, which will give you the angle cuts for common rafters. Keeping the square snug against the top edge of the board, tilt the square until the 6 on the common scale lines up with the same edge. Mark the angle on the board.

Step 3. Cut off the end, following the angled cut line. Measure 63 inches down the board, mark the bottom angle the same way, and cut it.

Step 4. Cut off the 1½ × 3-inch wedge from the bottom end of the rafter, as shown in the drawing below. Then measure 9 inches from the tail for the rafter and cut out the 1½ × 3-inch bird's-mouth notch, as shown below. Because a circular saw won't cut out the bird's mouth cleanly, use your handsaw to finish the cut.

Step 5. Once you have cut your rafter, use it as a template to make a second test rafter. Simply lay the cut rafter on another 2×6 and carefully redraw the cut lines and cutouts. You'll use these two as a test after you put up the ridge board.

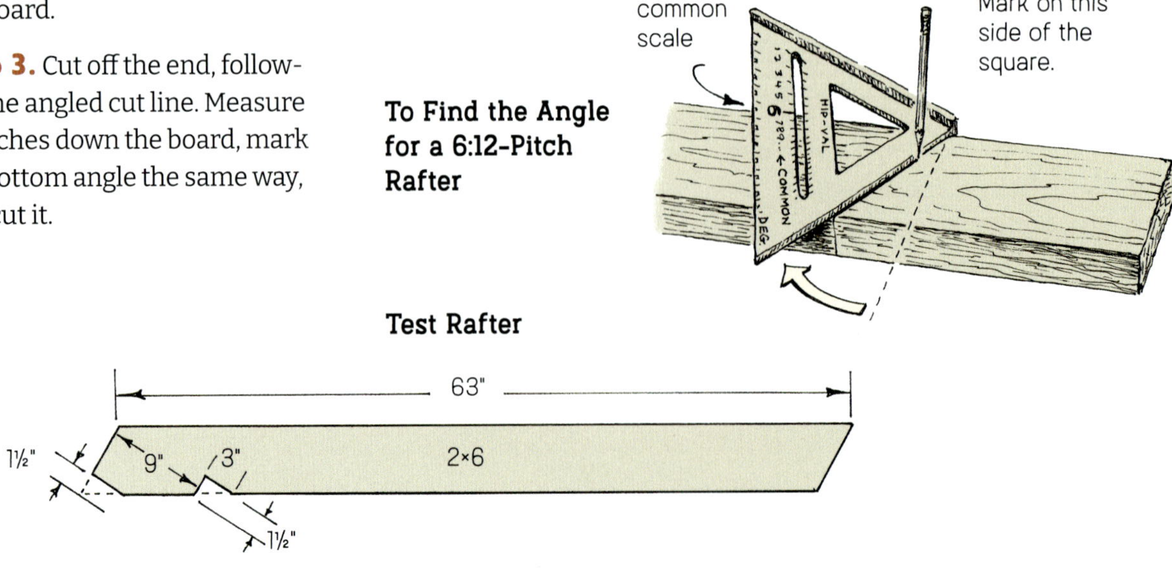

CUT AND MARK THE RIDGE BOARD

At the top of your roof, the ridge board helps line up all the rafters and supports them as well. To make your ridge board:

Step 1. Lay a 2×6 on your sawhorses and cut it to 17 feet long. This is the length of your building plus 12 inches on each end for an overhang.

Step 2. To mark it for the rafters, lay it flat. Measure and draw an edge line 12 inches in from one end, which represents the outside edge of your building. Draw an X on the *inside* of this line, which represents the position of the first rafter. Pound in a small nail at this edge line.

Step 3. Hook your tape measure on the nail and use it to mark edge lines every 16 inches. Use your square to draw the lines, then draw X's on the same side of the edge lines as you drew the first X. At the far end of the ridge board, you can omit the last rafter, leaving a 20-inch space.

Step 4. Draw another edge line 12 inches in from the far end of your ridge board. Draw an X on the opposite side of this edge line that you drew the rest of the X's.

Step 5. Using your square and pencil, transfer all the lines and X's onto the other side of the ridge board.

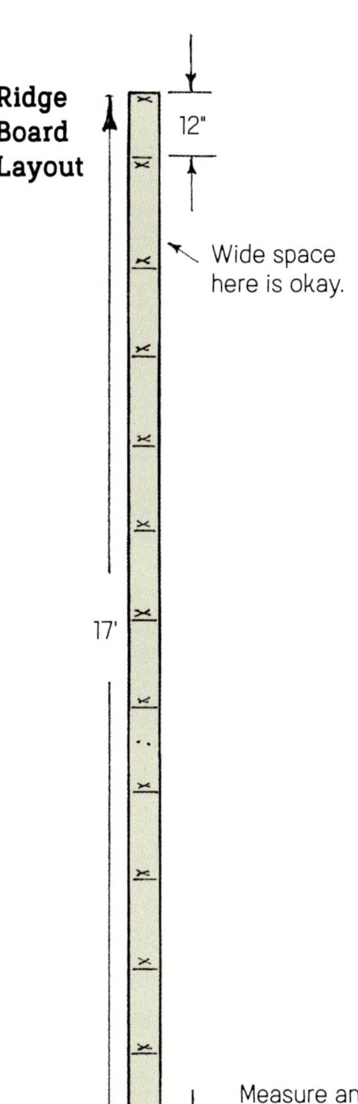

CUT AND MARK THE FASCIA BOARDS AND TOP PLATES

The fascia boards, which add strength and a finished look to the bottom of the rafters, are next.

Step 1. Cut two fascia boards the same length as the ridge board (17 feet).

Step 2. Lay the two fascia boards and the ridge board flat and side by side. Transfer all the edge lines and X's from the ridge board to one side of each of the fascia boards, then set the fascia boards aside.

Step 3. Next, use the ridge board to transfer the same rafter marks onto the long-wall plates. With the help of an assistant and some stepladders, hoist the ridge board onto the top of one of the long walls. Lay the board beside the long wall plate so the edge marks on the ridge board line up with the ends of the building. Carefully transfer all the lines and X's from the ridge board to the top of the plate.

Step 4. Move the ridge board directly across to the opposite long wall, and repeat.

Framing the Roof

To begin, it will be necessary to prop up the ridge board to hold it in the right position. You already determined that the rise of the roof is 2 feet, or 24 inches (see page 133). Theoretically, the bottom of the ridge board should be set at that distance—2 feet—to meet the ends of the rafters. However, the bird's-mouth notch will lower the rafters 1½ inches, so the actual distance from the top of the wall plates to the bottom of the ridge board becomes: 24"–1½"=22½". The next step is to support the ridge board to receive the rafters.

PROP UP THE RIDGE BOARD

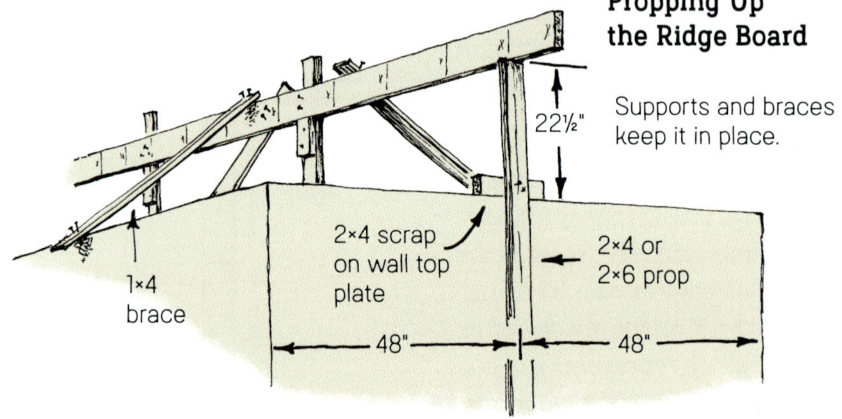

Propping Up the Ridge Board

Supports and braces keep it in place.

Step 1. Find the centers of the short walls, which are 48 inches in from the outside corners, and mark that spot on the top plate of each short wall.

Step 2. Get two 2×4s that are 8 feet long to use as prop boards, and two blocks or scraps of 2×4 or 1×6 about 1 foot long.

Step 3. Measure from the top end of each prop board to 22½ inches and draw an edge line. Using duplex nails, nail on the scrap pieces so their bottom edges are at the 22½-inch mark.

Step 4. Using 16d duplex nails, attach these prop boards to the centers of the end walls, resting the scrap piece on the top plate.

Step 5. Get two more 2×4s that are about 12 feet long, two more 2×4 scraps or blocks, and two or three 1×4s that are 6 to 8 feet long.

Step 6. Measure the distance from your floor to the top of the plates, then add 22½ inches to find the position of the middle support blocks. Measure out that distance on both long 2×4s, draw an edge line, then nail on the scrap pieces with duplex nails.

Step 7. With two stepladders and two helpers, lift the ridge board to the top of the end-wall props and set it there on edge. While your helpers hold each end steady, set the longer props under the ridge board somewhere in the center of the building and tack those in.

Step 8. Next, tack two or three 1×4 braces between the top of the ridge board and the top of a long wall, and one more long brace from the side of the ridge board to a short-wall stud near the floor.

Step 9. Sight along the ridge board with your eye. If it is sagging or bowed, adjust the support blocks and 1×4 braces until it is straight. Leave as many of these supports and brace boards in place as is possible until you've installed rafters.

TEST-FIT THE RAFTERS

Now it's time to test-fit the two rafters you've cut.

Step 1. With a helper, set the test rafters in place opposite each other, anywhere along the ridge board.

Step 2. Check the fit of the rafters. If the bird's mouths fit snugly over the plates, the tops of the rafters meet the top of the ridge board, and the angle cuts fit tightly, congratulations! If they don't fit right, check the measurements and the angles you cut. Trim these rafters or cut new ones, as necessary, until you have a good fit.

ridge board

48" 48"

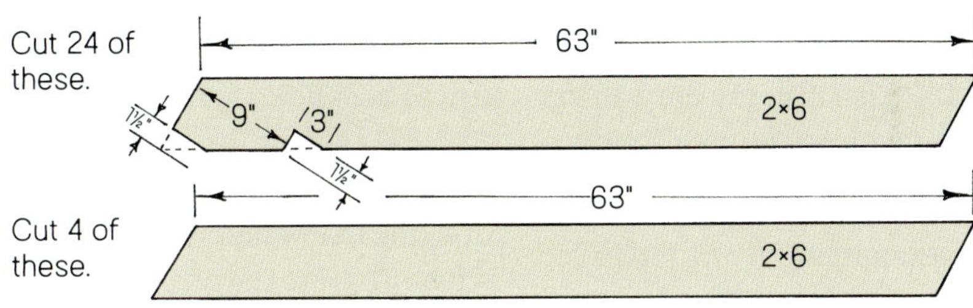

CUT THE RAFTERS

When your two test rafters fit, you're ready to cut the rest of the rafters.

Step 1. Write "template" on one of your good test rafters and use it to mark and cut the remaining 22 rafters the same way, for a total of 24.

Step 2. Cut the four *additional* rafters that will overhang the ends of the roof. Use the same rafter template for these, but don't cut out the bird's mouths or the small wedges at the bottoms. Set these four aside.

BRACE THE WALLS

You will need temporary wall ties to help keep the long walls parallel while you are putting up the rafters. Without them, the long walls will tend to spread out, and they will throw off your rafter fittings. (You can take down these temporary ties once you've installed the permanent braces, called collar ties; see page 138.)

Step 1. Cut two 2×4s (or 1×4s) to exactly 8 feet long. These will be the wall ties.

Step 2. Set the wall ties flat on top of your long walls, about 4 feet apart near the middle. Align the ends of the ties with the outer edge of the long-wall plates, keeping the ties clear of the X's that mark the rafter locations. Nail the ties into place with duplex nails.

INSTALL THE RAFTERS

First, nail in the rafters that attach to the tops of the long walls. You'll install the overhang rafters later, after you've installed the fascia boards. Set the rafters in place one at a time, and roughly in pairs to balance the weight against the ridge board.

Step 1. With a helper holding a rafter in place against the ridge board, toenail it into the wall top plates with 6d, 7d, or 8d nails.

Step 2. Attach the rafter to the ridge board by driving 16d nails through the ridge board into the end of the rafter, or toenailing the rafter into the ridge board, or both.

Step 3. Repeat for the rest of the rafters.

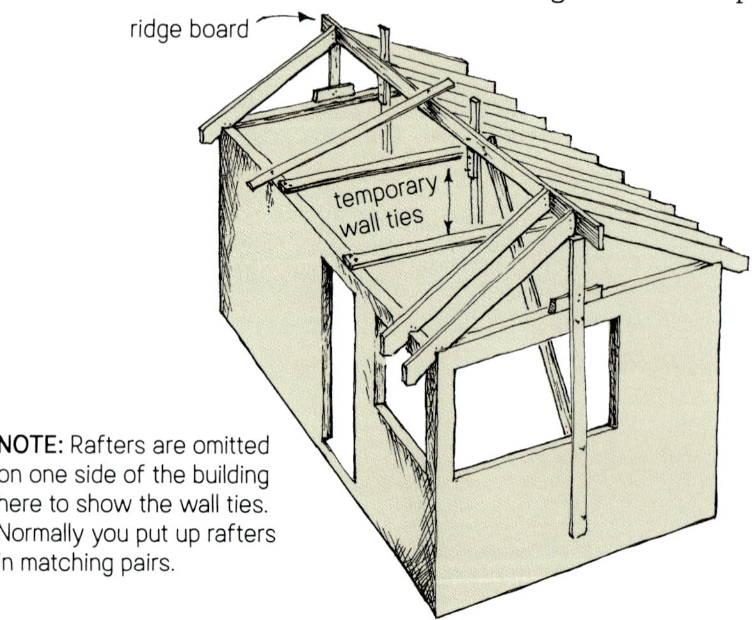

NOTE: Rafters are omitted on one side of the building here to show the wall ties. Normally you put up rafters in matching pairs.

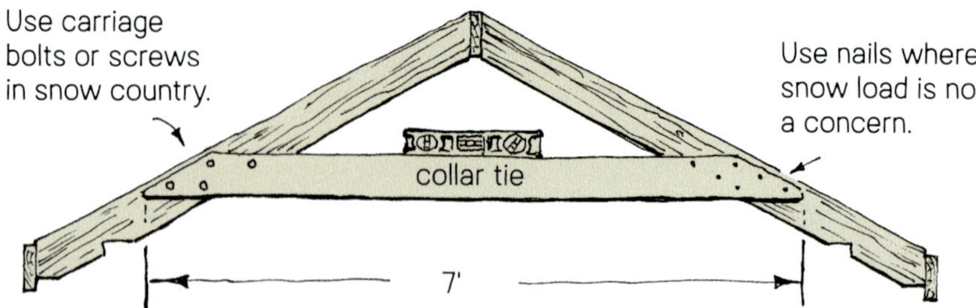

INSTALL THE COLLAR TIES

When all the rafters are nailed in, you will need to add collar ties to help hold up the roof. These should be installed across rafter pairs (see the drawing below) about every 4 feet. This means our 15-foot-long dwelling will require three collar ties. However, if you're building a loft, leave out the collar tie at that end, so you won't bang your head on it; the loft is designed to take its place.

Step 1. On your sawhorses, cut three (or two if you're installing a loft) 2×6s to 7 feet long.

Step 2. Cut off a 4 × 9-inch triangle from the upper corners of each 2×6.

Step 3. Start at least two 12d (not 16d) nails at each end of the collar tie. Drive the nails almost all the way through.

Step 4. With you and a helper at opposite ends, hold the collar tie to the pair of rafters, at about ¼ to ½ inch below the top edge of the rafters. Place your level on it, then both of you slide or nudge the tie until it's level.

Step 5. While your helper holds one end, take down the level and drive in the nails. Use five or six nails at each end. (If you live in snow country, see Fastening Collar Ties on page 139.)

Step 6. Once the collar ties are installed, remove the temporary wall ties.

Fastening Collar Ties

If you live in snow country, you may want to bolt or screw the ties to the rafters to handle a heavy snow load. Use 4-inch × ¼-inch carriage bolts or 3-inch construction screws. First, set up the tie to the rafters as before, again using your level. Tack the tie to the rafters with a 12d nail or two, or use a clamp to secure the tie to both rafters. For the bolts, get out your power drill, put in a ¼-inch bit, and drill four holes through the tie and rafters. Knock in the bolts with your hammer, add a washer and a nut to each, and tighten them with a wrench. For the screws, use at least six at each end of the tie. Drill pilot holes only through the first board, if necessary, and use your power drill to screw them in.

INSTALL THE FASCIA BOARDS

Here again, you'll need a helper and two stepladders.

Step 1. Put the two fascia boards you marked up earlier on your sawhorses. From the unmarked side or outside of each board, start a 16d galvanized box nail opposite the X for each rafter.

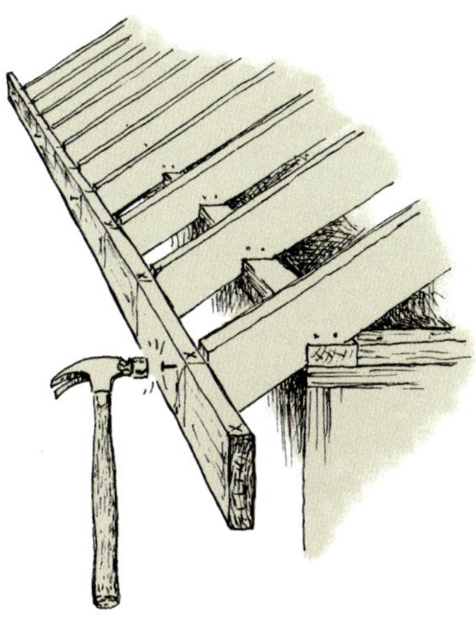

Step 2. With your helper holding one end, lift a fascia board into place. Check that the edge lines on the inside of the board line up with the rafter ends and that its top edge sits about ¾ inch below the top edges of the rafter tails, then nail the fascia board to the rafter tails. (Positioning the fascia board below the rafter tails ensures that the sheathing can lie flat against the rafters.) Start nailing at one end, pulling the fascia board up or down a bit as you proceed to align it properly with every rafter. Drive two or three nails through the fascia board into every rafter tail.

Step 3. Repeat with the other fascia board on the other long side of the roof.

INSTALL THE OVERHANG RAFTERS

Once the fascia boards are nailed on, you can put up those four overhang rafters, sometimes called barge boards, that you set aside earlier. If they fit snugly between the fascia board and the ridge board, they should automatically be in line with the other rafters.

Step 1. While your helper and you hold each barge board in place, drive in the nail you started on the fascia board into the bottom of the barge board.

Step 2. Toenail the upper end of the barge board into the ridge board with 6d or 8d nails. You're hanging out in space here a bit, so be patient.

Installing the Roofing

To lift the large panels of sheathing onto your roof, you'll need an extension ladder and a helper. Take your time with this until you get used to moving around on a sloped roof.

INSTALL THE SHEATHING

Step 1. Set a full sheet of ½-inch-thick plywood or OSB somewhere along one fascia board, aligning it so that any inner edges sit in the middle of rafters, leaving nailing room for other sheets of sheathing. You can set it at the end of the roof or in the middle. Tack it down with some 6d or 7d nails.

Step 2. Measure and cut other pieces of plywood or OSB to complete this first course of sheathing to the ends of the roof. Once they look good, nail on this first course of panels.

Step 3. Measure and cut the next course of sheathing to extend to the top of the ridge. Leave a 1½-inch gap between the ridge board and the top edge of the sheathing *except* where the sheathing extends beyond the walls (see the drawing at right). This will allow for ventilation in your roof. Stagger the seams in the courses of panels so they meet at different rafters, then nail them on.

Step 4. Repeat on the other side.

PUT ON THE ROOFING

For your roofing, you'll first lay down 15-pound felt roofing underlayment, generally called tar paper, then a metal drip edge all around the edges, and finally three-tab asphalt shingles in a color of your choice. You'll need a utility knife, a staple gun or hammer tacker, and tin snips, along with your usual tools.

Step 1. Unroll the 15-pound felt on the ground, measure to 18 feet, and cut off a piece with your utility knife. Repeat this until you have four pieces.

Note: Do not walk on the roofing felt until you've tacked it down—it might slide out from under you.

Step 2. Roll up one piece, carry it to the roof, and unroll it from one end along the bottom of your roof. Tack it down using a staple gun or, if you don't have a staple gun or hammer tacker, roofing nails. If you are going to shingle over it right away, you'll need only a few staples (or roofing nails) to hold down the underlayment. Use more on the exposed edges if it's windy. Trim off the excess felt at the gable ends with your utility knife.

Step 3. Repeat with the next piece, overlapping the first course by at least 4 inches. Cut away any felt that covers the ridge-top vent.

Step 4. Repeat on the other side of the roof.

Ridge-Top Venting

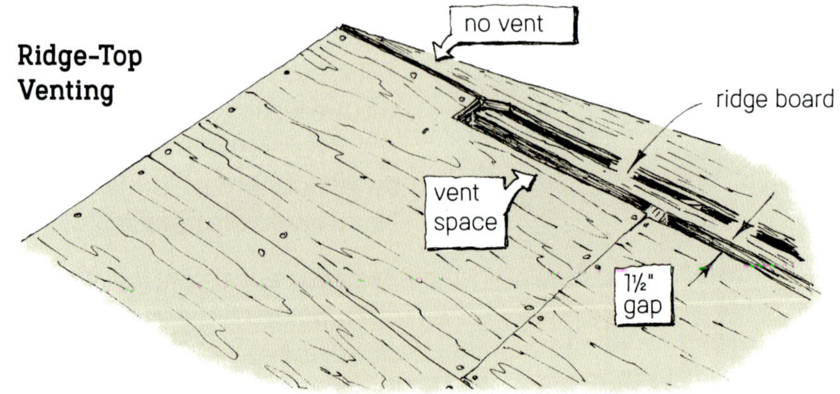

Extension Ladder Safety

Now is a good time to talk about using extension ladders. You'll need a sturdy, lightweight extension ladder to safely work on your roof. Set it on solid ground that won't sink or shift under the ladder feet. Lean it at about 15° (see the drawing below) and tie it to your wall or roof framing with rope or bungee cords if it's a windy day. Climb the ladder slowly at first until you get the feel of it. Trust the ladder and your sense of balance and you'll be fine.

Ladder Safety

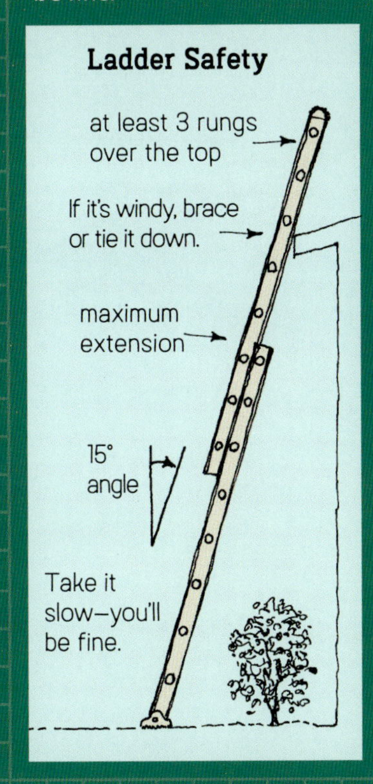

APPLY THE DRIP EDGE

Now it's time to apply the drip edge, which will protect your roof edges from the weather. (If you want to cover the end rafters with a wide trim board, though, which is nice-looking but optional, do that first.) It might be easiest to nail on the drip edge from a ladder, rather than from the roof. You'll need a pair of tin snips or heavy-duty kitchen shears to cut the drip edge.

Step 1. Place your first piece of drip edge on the lowest edge of the roof. Use roofing nails to nail through the top part of the drip edge.

Step 2. Working from the bottom to the top of the roof, continue to nail on the edging, overlapping the pieces by about 3 inches.

Step 3. When you reach the ridge top, cut the vertical part of the L-shaped drip edge at the ridge, then bend the top of the drip edge over the peak of the roof to get a nice-looking finish.

Roofing Felt and Drip Edges

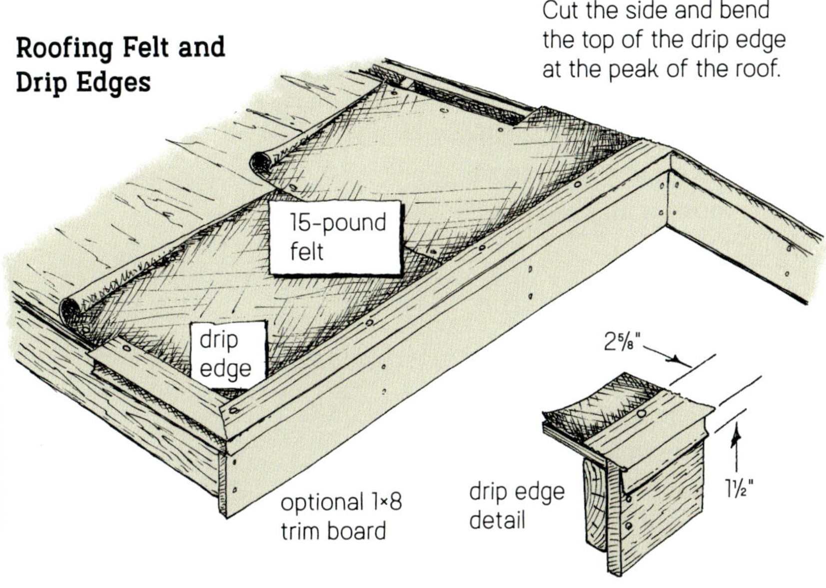

SHINGLE THE ROOF

Set a package of three-tab shingles on a plank on your sawhorses and read the directions on the package. You can easily cut asphalt roofing by scoring it on the back with your utility knife against a steel straightedge and then snapping the pieces apart. These are the basic steps for standard three-tab roofing shingles.

Step 1. For the starter course, cut off all the tabs from about five pieces of three-tab roofing. Save these tabs, because you can use them to cover the ridge of your roof. Nail the starter-course pieces to your roof, lined up on the edge of your drip edging.

Step 2. Stagger the next course by one-half of a shingle tab and set it to reveal a 5-inch exposure, which will have the tabs just covering the grooves of the course below. Drive two ¾-inch or 1-inch roofing nails 2 inches apart and about ¾ inch above each groove, and drive a single nail at the ends.

Step 3. Continue with your shingle courses, setting and nailing them in a staggered pattern. Trim the ends of each shingle course cleanly at the edge of your drip edges.

Step 4. When you get to the top, trim the shingles to keep the vent gap clear.

Step 5. Get your ridge venting and install it according to the directions. Different brands vary on how they are applied, but most allow shingle tabs to be applied over the venting for a finished look.

Step 6. To cover the ridge, use your saved shingle tabs, or cut three-tab shingles into single-tabs, and apply as shown.

How to Apply Three-Tab Roof Shingles

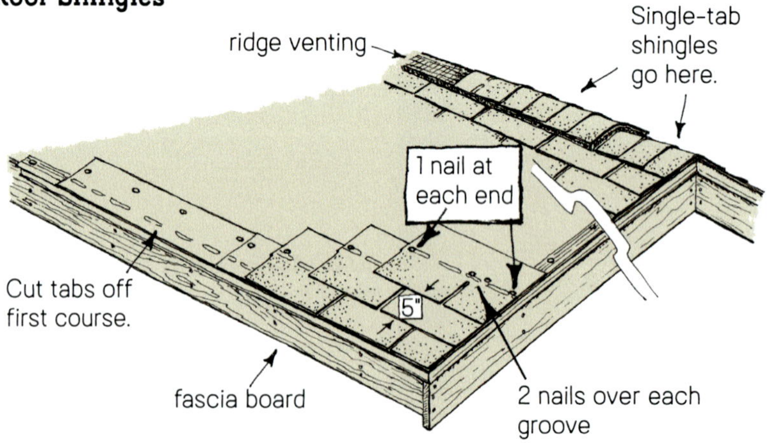

PUT IN A SOFFIT

With this roof design, you will need to cover the outside space under your roof, called the eaves, with a board called a soffit.

Step 1. Using tin snips, cut out 3-inch-wide strips of hardware cloth for vent mesh. Stretch these along the underside of your rafters and attach them with roofing nails. This is the lower vent for your roof, and the mesh will keep bugs out.

Step 2. Holding a small level to each rafter tail, make a mark on the wall across from the end of the rafter. Continue until you have a row of marks level with the bottoms of the rafters.

Step 3. Nail a 1×3 furring strip above the marks to hold your soffit board. Use a 6d or 8d nail every 12 inches.

Step 4. Use your 1×8 soffit boards to cover the underside of the overhang. Cut the boards so they join midway across a rafter tail, so each board has a solid surface to be nailed to. There should be a 1-inch- to 2-inch-wide gap exposing the wire mesh for the soffit venting.

Step 5. Nail up the soffit boards with 6d or 7d galvanized box nails. You'll be nailing upside down, which can be challenging. If you think it will be easier, you can use 1⅝-inch or 2-inch galvanized construction screws and your drill.

Putting in a Soffit

Trimming the Roof

For a clean look at the ends of your roof, you can cover the end rafters with 1×8 cedar or pine trim boards. If you decide to do this, put them up before you nail on the drip edge (see the drawing on page 141).

·CHAPTER 12·

Finishing Up

With the walls and roof framed and covered, it can rain all day and your place will still stay dry—until the wind picks up. So you need to finish closing it up. To do this, I'll show you how to build the walls in the gable ends and then put in the windows and the door.

Once your structure is closed up, you'll want to add the finishing touches to make the dwelling feel like it's truly your own. I'll talk about installing siding and trim for the outside, building a loft, and finishing the inside with wood or drywall. I'll also include a note about wiring your place for lighting and electrical outlets.

Finishing the Gable-End Walls

Get your stepladder and look at the still-open gable-end space under the rafters. Here you'll put in some short 2×4 studs extending from the top plate up to the rafters, then you can cover the remaining wall framing. Use leftover 2×4s and sheathing from your wall construction for these gable-end walls.

INSTALL STUDS IN THE GABLE-END WALLS

The studs in the gable space will line up directly above the studs in the wall below.

Step 1. Measuring from the wall corners, mark 16-inch centerlines, then edge lines, along the top plate.

Step 2. Next, use a level or plumb bob to mark edge lines for those same studs where they will attach to the rafters. Mark the edge lines on the rafter at the longer or higher edge of the stud location.

Step 3. Measure the distance from the edge line on the rafter down to the top of the plate to find the length for each stud, as shown on page 146.

Step 4. Cut the studs with angled notches to fit under the rafters. To find the angle cut for the notch, set your square on the edge of a 2×4, just as you did on the rafters—it's the same 6:12 pitch—and make the cut mark.

To cut out the notch, clamp down the 2×4 and saw the long leg of the notch with your power saw. Then with your handsaw, cut out the angled notch. Cut off the bottom or square end last. The unnotched leg remaining is 4 inches long for all the angled studs.

Step 5. Toenail the stud into the plate and into the rafter with 6d nails, or through the rafter with 12d or 16d nails. (These small studs bounce around, so use the nails that work best for you.)

Step 6. On the gable end of wall 3, arrange the studs around the window frame as shown on page 146.

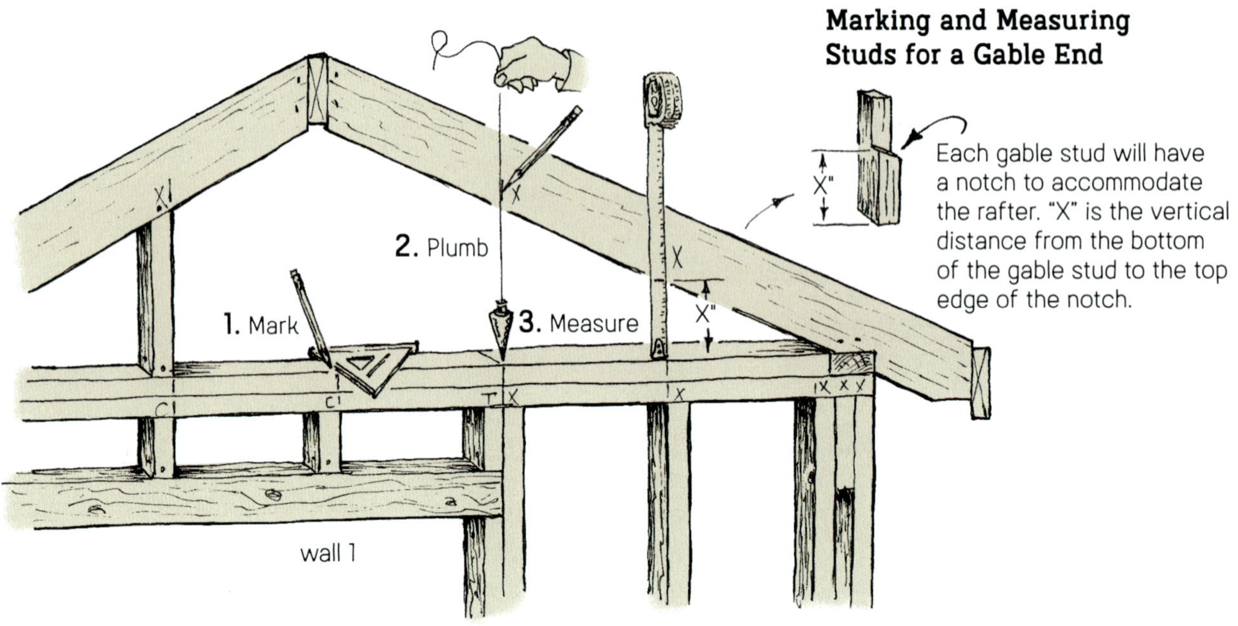

Marking and Measuring Studs for a Gable End

Each gable stud will have a notch to accommodate the rafter. "X" is the vertical distance from the bottom of the gable stud to the top edge of the notch.

Notching a Gable-End Stud

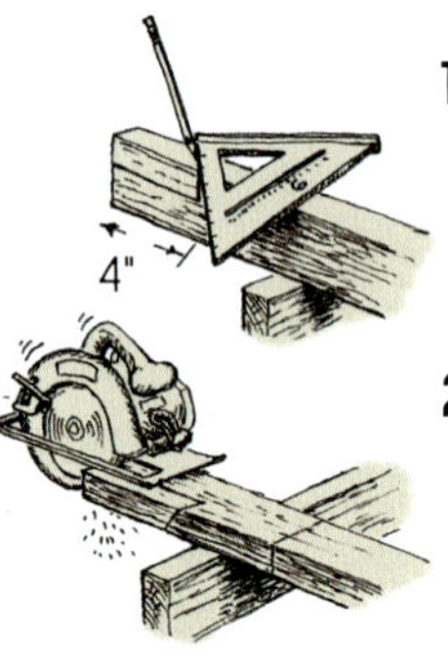

1. Mark the angle cut on the stud.

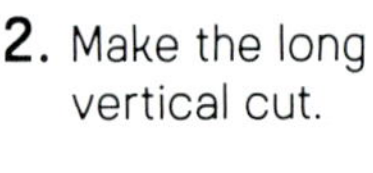

2. Make the long vertical cut.

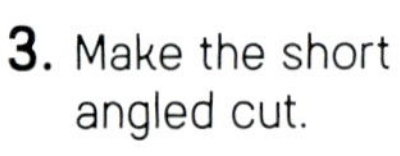

3. Make the short angled cut.

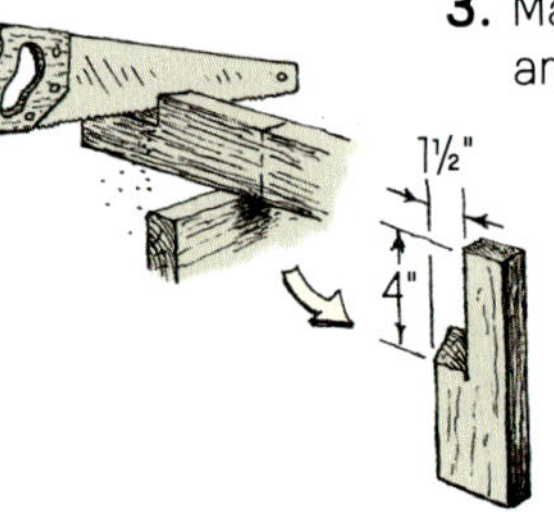

Wall 3 Gable End

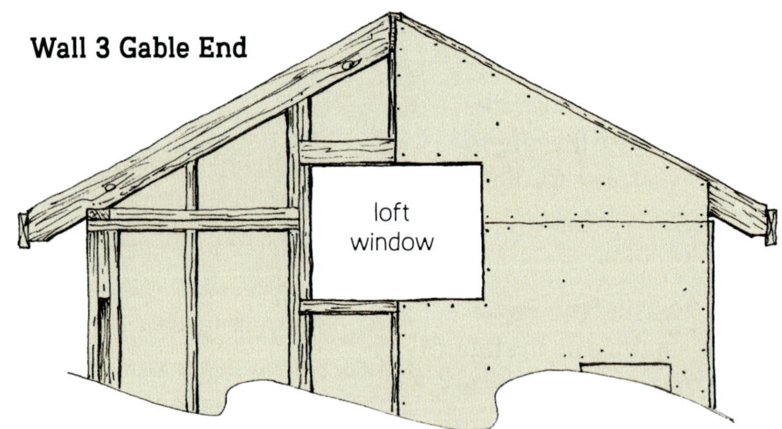

Now you can finish covering the gable-end walls with plywood or OSB. Cut sheathing pieces that will tie the window header and the rafter together. Use 7d or 8d coated sinkers to fasten the sheathing. For strong gable ends, especially over a wide window, apply construction adhesive to the stud edges before you nail on the sheathing.

Installing the Windows and Door

The right windows and doors can make an architectural gem out of the humblest structure. Conversely, with the wrong scale or size, they can make it look like an ungainly shed. Take some time to look around for the best choices for your design.

You can order new window units, such as horizontal sliders or casement windows, or you can build your own frames around recycled windows, barn sash, or custom-ordered windowpanes. In this example, we will build window frames to fit windows already on hand. The windows can then be hinged or fixed into the frames—it's your choice. You can also install your windows without the frames, as shown in Chapter 5.

New framed window units are relatively easy to put in, but you will still need to add trim boards around them for a good look. Unfortunately, off-the-shelf window units can be plain looking (i.e., boring) and scaled for bigger structures—like houses. Unless you're willing to spend a lot of money for custom-designed windows, I recommend building your own frames around old-style new or used windows.

One more alternative: You can custom-order relatively inexpensive and energy-efficient double-glazed windowpanes. Local glass-supply stores can have these glass-only (no frame) panels made for you to your exact size requirements. You would then install them as fixed windows without hinges.

DOOR AND WINDOW MATERIALS

PART	QUANTITY	DESCRIPTION
WINDOW FRAMES	10	1×6 boards, 8 feet or longer, "quality" or "no. 1" pine
EXTERIOR WINDOW TRIM	10	1×4 boards, 10 to 12 feet long, cedar or redwood
NAILS	2 pounds of each	8d finish nails; 6d or 8d galvanized box nails
DOOR	1	30 × 80-inch, or 2/6, prehung, left-hand, exterior, with lockset (a doorknob that locks with a key)
WINDOWS	varies	all openable windows will need two hinges and one latch each

As for a door, I recommend a quality exterior prehung, or already framed, door. A 30-inch-wide by 80-inch-tall door would be a good fit for your dwelling. It helps to have thought about how you'll finish your interior walls before buying a prehung exterior door.

At the lumberyard, ask for a prehung 2/6 exterior door. The 2/6 means the door itself is 2 feet 6 inches wide (30"). Ask for a frame or jamb that is 4⁹⁄₁₆ inches wide if you'll use ½-inch-thick drywall or 4¹³⁄₁₆ inches wide if you'll use ¾-inch-thick pine paneling inside. The lumberyard staff can then show you their catalog of door styles that is almost endless, so find a door that you really like!

Also, before you order a prehung door, determine which direction you'll want it to open: right-hand or left-hand. For a left-hand door, as you stand *outside* the door, the knob will be on the *right*, and the door will swing *away* from you, *to your left*. A right-hand door will do the reverse. You may have to explain this to the sales staff to get it right.

PUTTING IN THE WINDOWS

Take time to find the right windows. It is easy to walk into a large home center and be dazzled by the sheer quantity of windows on display. Most will not look right for your dwelling, though. Try to find a design that suits not only the size you need but also has some character, such as old-fashioned multipaned window styles, leaded windows, or even some with stained glass. The right windows will appear if you keep your eye out.

In this example, you will build and install box frames around the windows. Before making your box frames, however, measure the window openings and your windows to make sure everything will fit as planned. The wall openings are designed to be 1 inch bigger than the actual window all the way around. That allows for the ¾-inch-thick box frames and ¼ inch of wiggle room. Wiggle room allows space for adding shims to enable leveling of the window frame.

If the measurements look good, then you're ready to build the box frame.

It's Your Choice

As you've probably noticed by now, I tend to favor methods that give you the most value for your money. However, if you prefer guaranteed weather-tight and worry-free windows and you've found a custom design that you like, then by all means, spend what it takes and order something really nice. After all, you'll be living with your choice for many years! Install any factory-framed windows into your wall openings according to the manufacturer's instructions. Have someone help you set them in place—they can be heavy.

BUILD AND INSTALL A BOX WINDOW FRAME

You'll cut the box-frame pieces from 1×6 boards. In addition to being cut to length to form the box frame, the 1×6 boards will have to be ripped, or cut lengthwise, to fit into your walls nicely. The window frames should be 4⁹⁄₁₆ inches wide if you'll use ½-inch-thick drywall inside, or 4¹³⁄₁₆ inches wide if you'll use ¾-inch-thick pine paneling inside.

Step 1. Cut the 1×6 boards to the lengths and widths you need for your box frame (see page 154 if you need advice on using a rip fence with a power saw).

Step 2. Assemble the frame, as shown in the drawing at right. Use three or four 6d or 8d box nails per corner.

Step 3. Slide the frame into the wall opening and set it with shims so it will stay put. Using a level, adjust the shims as necessary to make the window frame both level and plumb.

Step 4. Adjust the outer edge of the frame to make it even or flush with the outside surface of the sheathing. When it's good, nail it in with 8d finish nails, driving the nails at the shim locations. (If you nail where there are no shims to fill the gap, the frame board will bow toward the framing.)

Step 5. Apply 1×4 trim to the outside of the window. Set these trim boards to overlap the inside of the frame edge by ½ inch to hold the window in place (see the drawing below right). Nail the trim with 6d galvanized nails into the edge of the window frame, and with 6d or 8d galvanized box nails firmly into the wall around the window. Alternatively, you could nail the trim boards onto the frame on your sawhorses and then put the entire assembly into the wall.

Step 6. Caulk the top of the top trim board, or use Z flashing, as shown below right.

Box Frame for a Window

Installing a Window Frame

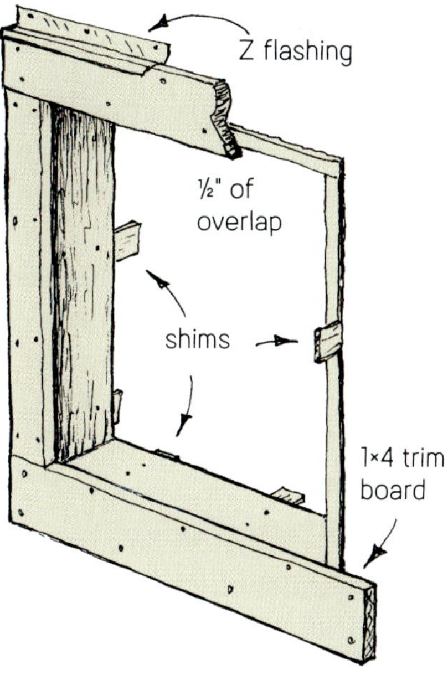

INSTALL THE WINDOW

If the window will be fixed (no hinges), fit it in the frame, and nail a 1×2 piece of trim or furring strip into the frame all around the inside of the window. Outside, caulk the window all around where it meets the exterior trim boards. If the window will be openable, follow these steps:

Step 1. Set the window on your sawhorses and attach a thin metal rain shield to its outside bottom edge. You can make this shield by trimming and then bending a flat strip of aluminum flashing, which is sold in rolls and used for weather protection. Use tin snips or kitchen shears to cut it. You could also use a piece of Z flashing, with its bottom part bent up.

Step 2. Test the length to get as close of a fit as possible, then attach the rain shield to your window frame by gently pounding in 1-inch-long copper brads (tiny nails), or use round-headed wood screws. Caulk or seal the top edge of the metal to the wood. This will help keep out the rain.

Step 3. Set the window into the frame and shim it so it is plumb and level and doesn't scrape the bottom of the frame.

Step 4. Attach the hinges, following the instructions on pages 68 and 69.

Step 5. Finally, attach a twist latch (called a casement latch) to the window, which will close it tightly.

Window Rain Shield

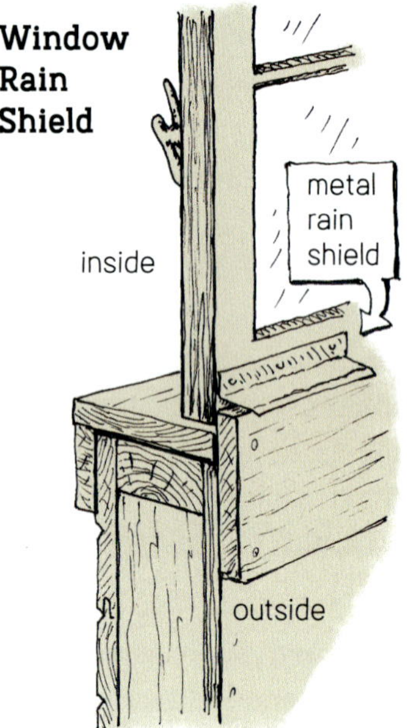

Finished Window

half-opening, half-fixed double window

PUTTING IN THE DOOR

Don't put the cart before the horse. Make sure you have the door on hand, or its specific rough-opening requirements, before you frame the wall opening. This project uses the standard rough-opening size (called RO in carpenter-speak) of 2 inches wider and 2½ inches higher than the actual door size, but some prehung doors may require a different rough opening.

Your prehung door will probably arrive all wrapped up. First, check to see if it is the size you ordered and that it opens in the direction you asked for. Find the installation instructions and read them carefully. You'll likely be installing the doorframe together with the door into the wall opening. Have an assistant help you set the door in place. Use shims to set it plumb and level, and tack it in place with nails, as instructed by the manufacturer. Also check to see that it is flush with the outside sheathing surface.

Prehung exterior doors often include an outside trim board, usually called "brick molding." This will work fine, but you can also replace it with your 1×4 or other trim if you prefer. Most prehung doors come with holes drilled for the doorknob, or lockset, and many recommend the brand of lockset that is compatible with the drilled holes. This is good. Choose a style of lockset that you like for your door, and install it per the instructions.

Installing a Prehung Door

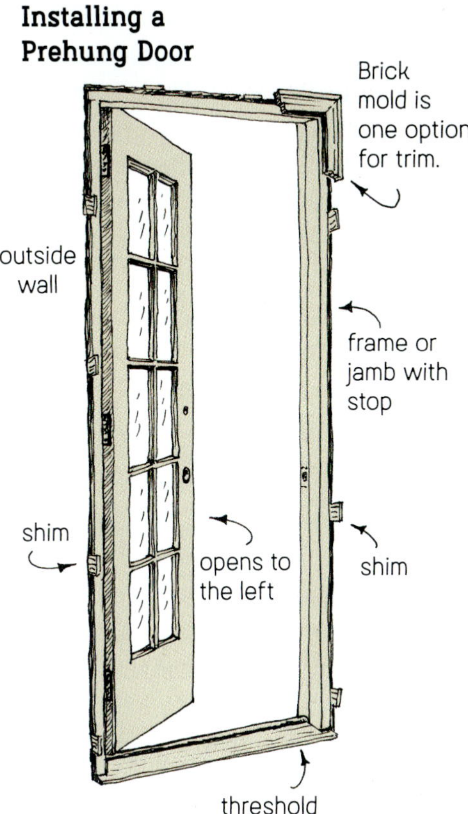

HOW TO BUILD YOUR OWN DOORFRAME

If you have found the perfect recycled door and want to build a frame or jamb for it—like the window frame just described—or if you've purchased an exterior doorframe kit, follow these steps for putting the pieces together and installing the frame in your wall.

Step 1. Check your door to make sure it is square. Some old doors have had their tops or bottoms shaved off at angles to fit sloping floors. You may have to resaw the top or bottom with your power saw to square it.

Step 2. So that it will be flush with both the outside edge of the exterior sheathing and the inside finished wall, your doorframe should be 4⁹⁄₁₆ inches wide if you'll use ½-inch-thick drywall or 4¹³⁄₁₆ inches wide if you'll use ¾-inch-thick pine paneling inside. If you're working with a kit, look for one in the appropriate width. If you're working without a kit, you'll need to rip 1×6 lumber to the appropriate width. Do that now; see page 154 for instructions on using a rip fence with a power saw. Cut pieces for the top and two sides of the frame. The bottom piece will be the threshold, a sturdy, slightly sloped board that fits just under the door and helps shed rain; you can buy wooden exterior thresholds at most lumberyards.

Step 3. Once you've checked your door and made it square, cut the top piece of your frame long enough to allow for the width of the door, the thickness of the two side pieces of the frame, and ⅛ inch of wiggle room. Cut the threshold just long enough to allow for the width of the door plus ⅛ inch; don't include the thickness of the side pieces in its length.

Step 4. Nail the pieces together, as shown in the drawing on the facing page. If the top piece and threshold of your doorframe is built with a doorstop as one piece, trim the ends off the doorstop before assembling the frame to accommodate the side pieces. Saw any extra wood off the doorframe that extends below the underside of the threshold.

Step 5. Set the doorframe in your rough opening, with its outside edge flush with the outside sheathing surface. If you bought an exterior doorframe kit, check that the doorstop edge is facing outside, so that your door will open to the inside.

Step 6. The doorframe should fit loosely in your rough opening, with the threshold resting on the floor. Use your level to make it square, then use shims and a nail or two to hold it in place for nailing. Test the door by setting it in the frame. If it looks good, take out the door, then shim and nail the frame in the opening as with a window frame. Use two 8d finish nails every 16 inches or so.

Step 7. Get a set of three 3- or 3½-inch butt hinges, with removable pins, which are made for doors. The hinges will need to be mortised, or set, into the edge of the door and into the frame, and they must be lined up for the door to hang properly. Use a hinge to mark its edges on the door for all the hinges. Take out the pins and separate the hinge halves.

Step 8. Set the door on edge on the floor and chisel out the hinge mortises just deep enough so the hinge surface is level with the wood surface around it—go no deeper. Then mark the screw holes with the hinge half. Screw the hinge halves into the door, but not tightly.

Step 9. Set the door in the frame, using thin shims at the bottom to create a small gap. Mark the hinge edges onto the doorframe.

Remove the door. Carve out the space for the hinges as you did on the door, and screw the hinge halves into the frame, but not tightly.

Step 10. Hang the door by joining the hinge halves. Tap the pins into the hinges and tighten up all the screws. This always takes more time than I expect, but it works!

Step 11. If your old door doesn't already have a doorknob or lockset, install your doorknob hardware according to the instructions, if available. Chisel out a striker-plate hole for the door latch into the frame and install the striker plate.

Step 12. Add a doorstop (but skip this step if you used a doorframe kit with a doorstop included); you can find pine doorstops at most lumberyards. Cut the strips to size. Close the door gently so it is latched, then draw a line all around the doorframe to determine the edge of your doorstop. Using 4d finish nails, nail the doorstop, which will also help provide a weather seal around the door, into your frame. All this can be done after the door is hung.

All this carving, fitting, and futzing can get a bit frustrating, so be patient, turn up some music, and try to go with the flow.

Exterior Door Frame Assembly

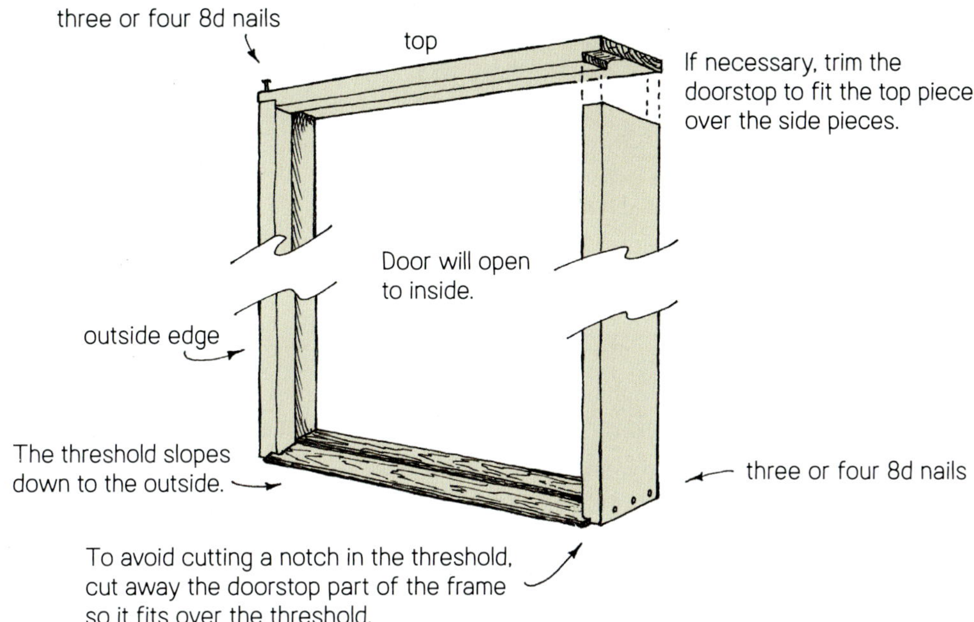

Building a Loft

Now it's time to build the loft. The loft will also help tie the walls together, so if you decide *not* to build it, add one more collar tie to your rafters (see page 138). You will need your usual tools, a step-ladder, and these materials.

LOFT MATERIALS

PART	QUANTITY	DESCRIPTION
LOFT BEAMS	3	2×6s, 8 feet long
LOFT JOISTS	7	2×4s, 6 feet long
LOFT FLOOR	9	1×8 car siding boards, 8 feet long; *or*
	12	1×6 car siding boards, 8 feet long
SCREWS	1 pound	2½-inch construction screws
NAILS	1 pound of each	16d coated sinkers; 8d finish nails

LADDER AND RAILING MATERIALS

PART	QUANTITY	DESCRIPTION
LADDER RAILS	2	2×6s, 8 feet long
LADDER STEPS	1	5/4×6 cedar decking board, 10 feet long (to be cut into five steps, each 21½ inches long
RAILING	1	5/4×6 cedar decking board, 6 feet long
SPINDLES	3	pine 2×2s, 8 feet long (to be cut into twelve 23½-inch railing spindles)

CUT THE JOIST SUPPORT PIECES

The loft frame and floor will be supported by two beams. Each beam is made of a 2×6 and a joist support piece, joined in an L formation. The rear beam will be 7 feet 5 inches (89"), and the front beam will be 8 feet (though you can shorten these lengths a bit if the beams fit too tightly). You'll rip the joist support pieces from an 8-foot-long 2×6 using a power saw and a rip fence.

Step 1. The two joist support pieces are exactly 2 inches wide. So start by using your tape measure to set the rip fence blade 2 inches from the face of the saw blade, then tighten the fence screw. Draw a short cut line 2 inches from the edge of the 2×6 to confirm the rip-fence position when you start cutting.

Step 2. Clamp an 8-foot-long 2×6 to your sawhorses. The saw will want to tip over while you are cutting, so set another 2×6 or a 2×4 over your sawhorses to help support the saw.

Step 3. Begin the cut. Guide the saw slowly as you get used to keeping the fence snug to the edge of the board. If it wanders, turn off the saw, back it up, and then recut it straight.

Step 4. Flip the cut 2×6 over so that you can run your saw along the remaining straight factory-cut side and cut another 2-inch-wide piece from the board. Now you have two joist support pieces.

Rip-Sawing a 2×4

INSTALL THE FRONT AND REAR BEAMS

For a stronger building, the loft beams will be firmly attached to the wall framing. The outer edge of the front beam will extend to 4 feet 9 inches (57") from the rear wall.

Step 1. Measure from the rear wall along both side walls to 57 inches and mark the floor at each spot.

Step 2. On both side walls, the mark should fall between a pair of studs. Between these studs, attach a 2×6 block to support the loft beam, setting it so that its top is 66 inches up from the floor. (For extra strength, you can add a support stud under it, which should be about 59 inches long.)

Attaching Loft Beam

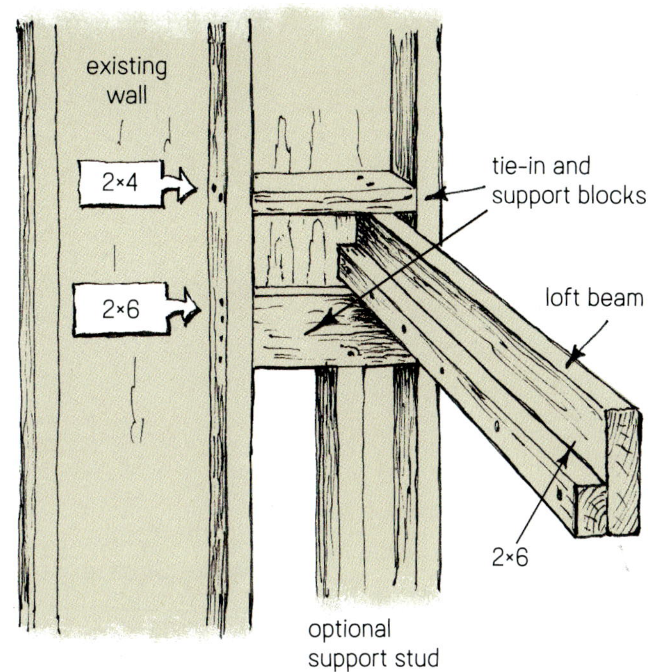

Step 3. Measure again from the end wall to exactly 57 inches and draw a line across the top of each 2×6 block at this spot. These lines mark the position of the outside edge of the front beam. Set the beam in place, aligning its outer edge with the lines you just drew. Then add another 2×4 block, laid flat, on top of the beam to better tie it to the wall.

Step 4. Attach one of the 2-inch-wide joist support pieces to the inside face of the front beam. I recommend using 2½-inch construction screws, set about 12 inches apart. For easier screwing, first drill ⅛-inch pilot holes through the 2-inch-wide joist support piece.

Step 5. On the rear wall, measure 66 inches up from the floor and draw a horizontal line across the studs. Tack a couple of 16d nails on this line, then rest the 2×6 rear beam on these nails to hold it in the right place. Nail or screw the 2×6 directly into all the studs it meets. Use at least two 16d nails or 2½-inch screws per stud.

Step 6. Attach the remaining 2-inch-wide ripped joist support piece to this beam, as you did for the front beam.

INSTALL THE JOISTS

Step 1. Measure and cut seven 2×4s to fit snugly between the loft beams, resting on the joist support pieces between the beams.

Step 2. Set the two end joists in place and fasten them by driving 16d nails directly through the joists into the studs on the side walls. Then set the remaining joists in place, spacing them evenly across the loft floor frame (at about 14¾ inches center to center).

Step 3. For a clean look, without nailheads exposed on the outside of the front loft beam, toenail each joist with only one 6d or 8d nail through the top of the joist and into the beam. The 2-inch-wide support pieces will otherwise hold the joists in place.

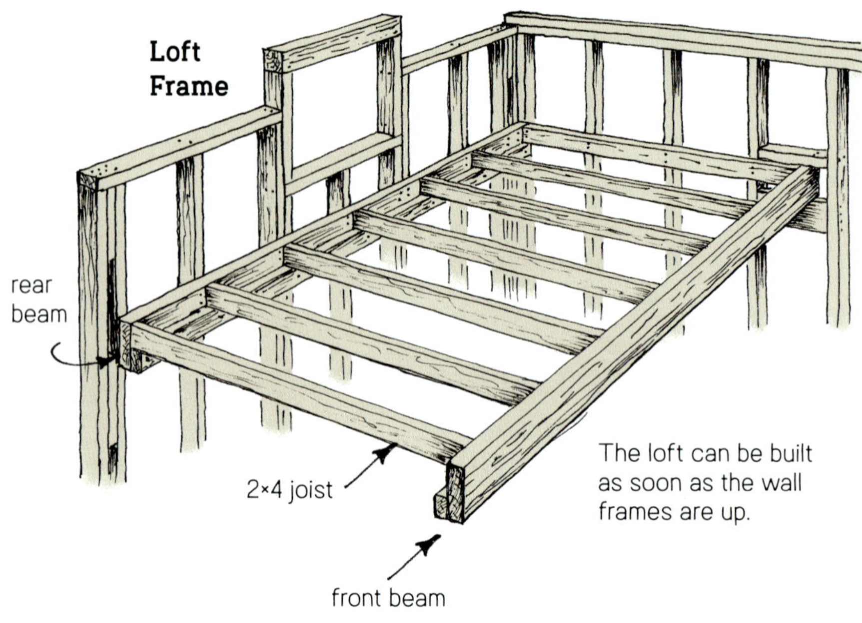

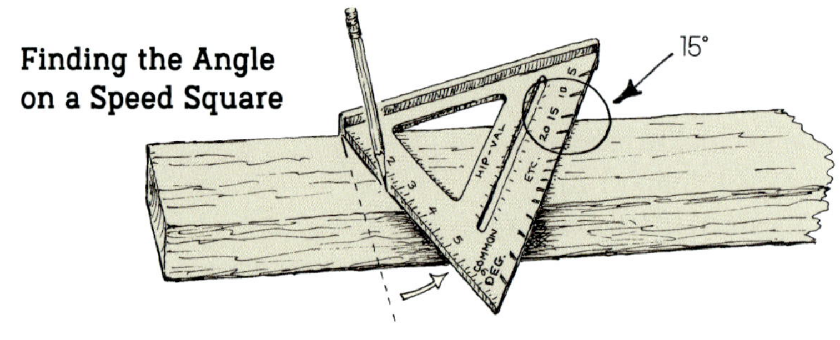

ADD THE FLOOR, RAILINGS, AND ACCESS LADDER

Step 1. For your loft floor, lay down car siding, with the grooves facing down, or some other ¾-inch-thick pine boards. Start at the back wall and work forward toward the loft edge. Use 8d finish nails, two or three per joist, and use your nail set to avoid dinging the loft floor with your hammer. When you get to the last board, you'll probably need to rip it to width (as you did for the loft joist supports; see page 154). After nailing it in place, plane or sand its edge for a smooth, finished look.

Step 2. Next, build a ladder so you can get up to your loft. You can build an access ladder with rungs, but for a more comfortable ladder, build it with flat steps instead, as shown on page 157.

Step 3. Finally, build a railing by attaching spindles every 4 inches to a top rail of 1-inch-thick cedar decking. With a helper, attach the assembled railing to the loft beam with two 2½-inch-long screws per spindle.

Note: To prevent splitting, pre-drill the screw holes before driving the screws through the spindles.

Loft Ladder and Railing

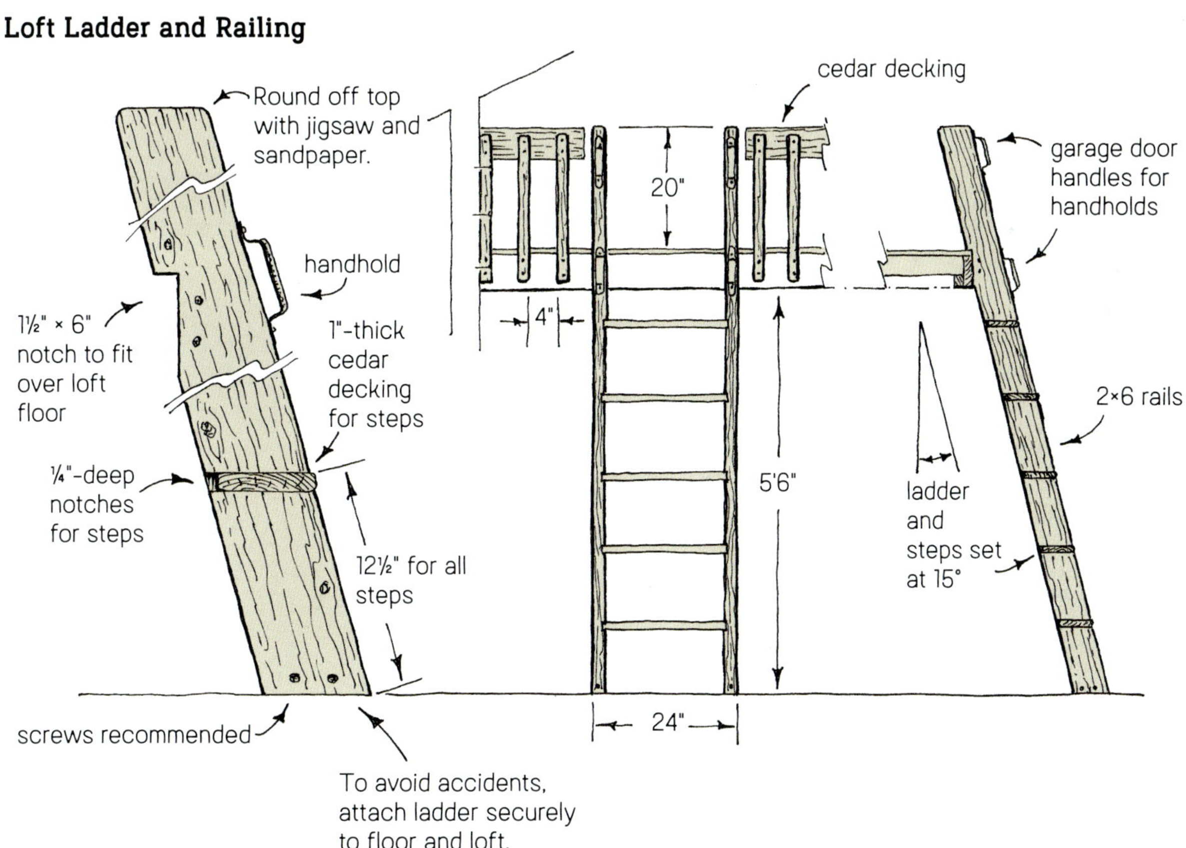

Installing Trim and Siding

Take some time to think about the exterior look of your backyard dwelling. Do you want it to match a bigger house nearby or to fit your own vision of a personal space? Traditional wood siding is covered extensively in Chapter 7, but other materials such as stucco and unique items such as recycled doors are also worth considering. The trim materials listed below will work with the traditional siding materials.

TRIM AND SIDING MATERIALS

PART	QUANTITY	DESCRIPTION
WEATHER BARRIER	400 square feet	housewrap or 15-pound felt
TRIM	12	1×4s, 12 feet long, cedar or redwood
TRIM	3	1×6s, 12 feet long, cedar or redwood (optional; see trim drawing on the facing page)
SIDING	360 square feet	your choice of material and style
NAILS	1 pound	7d or 8d galvanized box nails (for trim)
NAILS	5 pounds	6d or 8d galvanized box or siding nails (for siding)
SEALANT	2 tubes	paintable acrylic or silicone exterior caulk

WRAPPING THE WALLS

Before installing the trim and siding, wrap your dwelling with a semipermeable, protective layer of plastic. Commonly called housewrap, this material comes in wide rolls and keeps rain out while releasing moisture from inside the house. For wood siding, ask for the ribbed design, which doesn't trap moisture behind the boards as much as the smooth variety. Or use 15-pound felt roofing underlayment (tar paper) for the same purpose. Have someone help you put up the wrap, then tack it on using a staple gun.

PUTTING UP THE TRIM

A good trim material is 1×4 cedar. Cedar cuts easily, won't shrink or crack as much as pine, and will hold up to the weather for a long time. Use your handsaw to cut it if you find the circular saw strays too much from your cut line. Nail it on with 7d or 8d galvanized box nails.

Take the time to fit the wall trim boards carefully, especially at the gable ends. Cut the angles under the gable-end roof at the same 6:12 pitch as you did the rafter ends. The drawing below suggests wide trim boards under the window and roof overhang, but feel free to alter the width any way you'd like. Use pieces of cedar or exterior knot-free plywood to cover the soffit ends.

PUTTING UP THE SIDING

See Chapter 7 for a thorough discussion of siding options. If you find someone replacing their pine or cedar siding with vinyl siding, ask if you can salvage their old boards. They will likely be delighted to give them to you to save dumpster costs. Salvage yards and Habitat ReStores might also have recycled wood siding. You may have to remove some loose paint, but a possible unique look, and the low cost, may make this worth it.

For good weather protection, caulk any minor gaps between boards and the joints where the siding meets the edges of trim boards, especially around the windows and door. Acrylic or silicone caulk with long-lasting flexibility is best. Quality cedar shingles and trim can take the weather without being painted, but other raw wood siding should be stained or painted after the caulking is completed.

Gable-End Trim

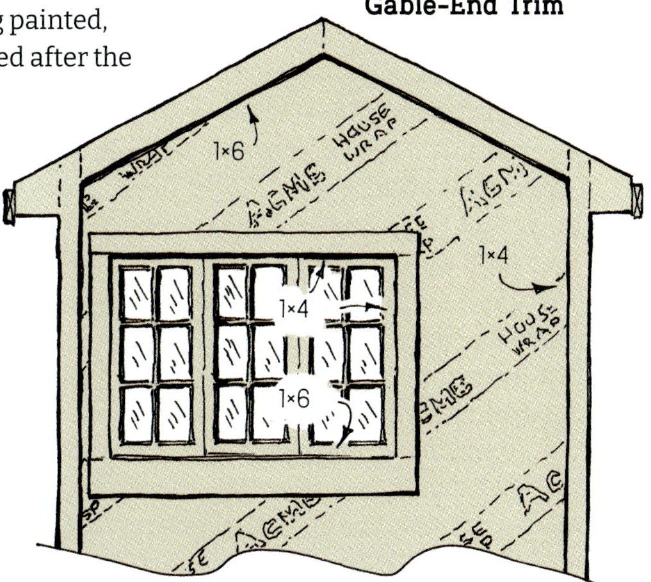

Installing Plumbing, Wiring, and Heating

The next step is to decide on your electrical, plumbing, and heating needs. If you want a sink or a toilet, this will mean installing plumbing. To get water, you can run an underground pipe from your house, or even use a garden hose to connect to your household water source. If you want hot water, you can add an electric flash heater to your water line. In cold climates, you will need to protect any outside water pipes from freezing. The sink can drain out to a dry well (if allowed in your area), which is essentially a buried container of gravel that helps the ground absorb your sink's wastewater.

I suggest a portable toilet for bathroom needs. Common on boats and RVs, this self-contained device can be periodically emptied into a regular toilet. A small composting toilet vented to the outside may also work, and there are many types of these toilets to choose from.

Will you want electricity? If so, for what? If you are thinking about food, you will need outlets for a microwave and mini-fridge. Will you want a reading lamp by the bed and power for your computer, entertainment system, or a classic lava lamp collection? Will you have lights in the ceiling or outside? Will you want cable or internet?

Electrical installation will mean running wire through your walls. In houses, the electrical power comes from the utility company to a breaker panel, or breaker box. In your backyard dwelling, the power will likely come from the main breaker panel in your house to a smaller breaker panel, or subpanel, in the dwelling. Any plumbing pipes for a sink or toilet will also be run through the walls and will drain through the floor. In general, when it comes to electrical and plumbing, first check with the local building department for all applicable permitting requirements in your area. In most jurisdictions, though, you can do most—if not all—of the work yourself, since this is your place. I recommend finding an experienced electrician and plumber to help you design and install your wiring and plumbing. Once you've seen it done, it will be easy, but there are safety and code requirements to consider.

The final consideration before you finish your walls is heating and air conditioning. For a structure this size, a small electric heater will most efficiently do the job. Baseboard heaters are also available, but most of these require 220/240-volt wiring. Another option is a "mini-split"—a combined heating and air-conditioning unit that mounts on a wall. If you prefer a small wood-burning heater or fireplace, determine how you will install it, including a chimney and surrounding tilework or other heat barrier, before you finish your walls.

Once your wiring, outlet and light boxes, plumbing, and heating/air conditioning are roughed in, you can finish your interior.

Adding Insulation

If you want to keep your dwelling warm and cozy when it's cold out and keep it cool when it's hot out, consider insulation. The roof of this project has been designed to accept R-11 or R-13 fiberglass insulation, which will allow 2 inches of free air space for ventilation. R-19 (6×) insulation will fill up all of your vent space, which may cause mold or rot to form inside your roof. Since you set the rafters on 16-inch centers, get the 14½-inch-wide rolls. Use the same R-11 rolls for the walls.

For the walls and ceiling, cut the insulation strips with your utility knife to the length you need. Fill all the spaces, including the narrow gaps between close-together studs. Use a staple gun to attach the reinforced paper edges of the insulation to the outer edges or sides of your studs and rafters (see the drawing at right).

For the final insulating step—and this one is optional—stretch thin polyurethane sheeting over your walls and ceiling and staple it down. This will help keep moisture from sweating inside your walls during cold weather and hold in a bit more heat.

Fiberglass Insulation

Finishing Your Walls and Ceiling

For the interior finish, I suggest using pine paneling or a combination of paneling and drywall. Tongue-and-groove pine car siding, or knotty pine, depending on the style, is a beautiful and inexpensive natural wood that you can stain, varnish, or paint. It is best to apply pine boards horizontally (across the studs) on the walls or ceiling. A balance of pine paneling and painted drywall can create a natural yet restful look.

INSTALLING PINE PANELING

To put up 1×6 or 1×8 pine paneling such as car siding, cut the ends cleanly by using your square and a utility knife to score the cut line instead of drawing it with a pencil. Saw it with a sharp toolbox saw. For ripping, or cutting a board lengthwise, use your power saw and rip fence as shown on page 154. Nail in the boards with 8d finish nails, then—to avoid hammer dimples—set each nail with your nail set.

HANGING DRYWALL

The interior wall covering used by most house builders is drywall, which is also called plasterboard, wallboard, Sheetrock, or gypsum board. It is inexpensive and durable, and it can give a small room an elegant look. The standard 4 × 8-foot × ½-inch-thick drywall sheets are heavy, so find an assistant to help you put them up. Finishing or plastering drywall can be challenging, but it just takes practice. Be sure to store drywall out of the weather—rain will ruin it.

Drywall Tools and Supplies

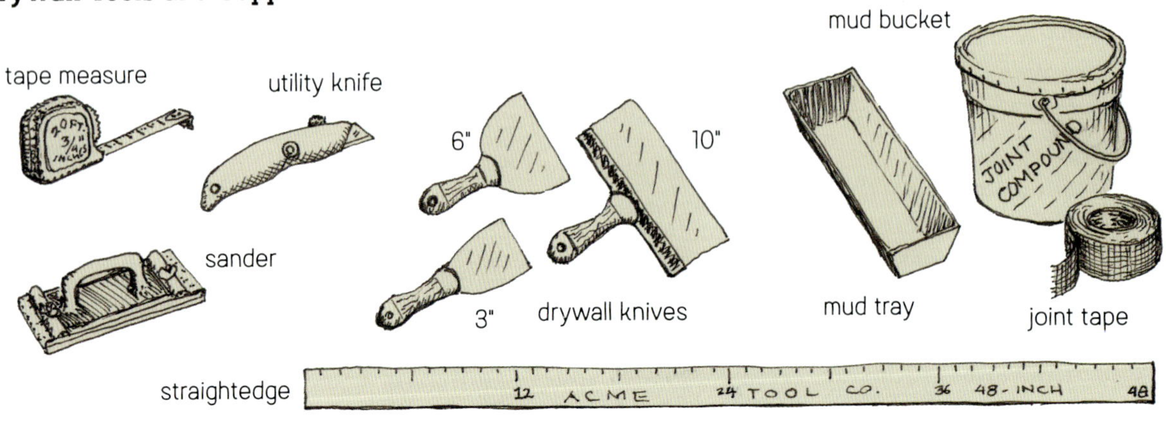

tape measure
utility knife
6"
10"
mud bucket
JOINT COMPOUND
sander
3"
drywall knives
mud tray
joint tape
straightedge
12 ACME 24 TOOL CO. 36 48-INCH 48

HOW TO CUT DRYWALL

You will need your tape measure, pencil, utility knife, and a straight board or steel straightedge at least 48 inches long.

Step 1. Measure and draw your cut line, using the straightedge. Cut through the surface paper with your knife along the cut line.

Step 2. Snap the sheet at the cut.

Step 3. Turn the sheet over and cut through the back layer of paper.

How to Cut Drywall

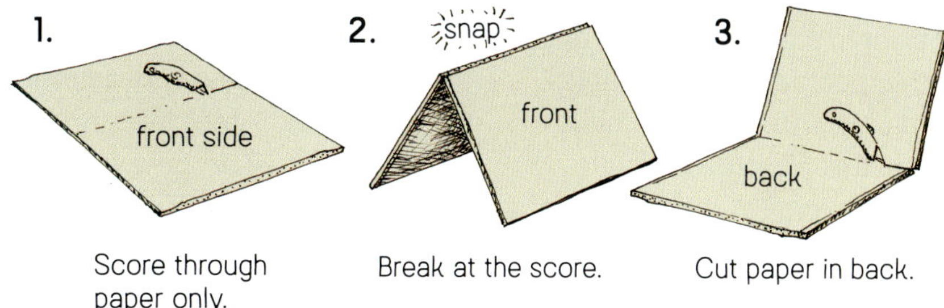

Score through paper only.

Break at the score.

Cut paper in back.

HOW TO HANG DRYWALL

Having a helper makes a big difference, especially on the ceiling. Use 1¼-inch drywall screws or nails for the walls, and 1⅝-inch drywall screws or nails for the ceiling (I recommend screws). For easy application of drywall screws, get a special drywall screw "dimpler bit" for your power drill. This inexpensive wonder automatically sets the screws to the correct depth, making the job easier.

Hang the large pieces of drywall first to minimize cutting, then use the scraps to fill in the smaller places.

Step 1. Set a piece of drywall in place. Make sure all the edges of the piece will sit over studs or other solid framing, leaving room for attaching other pieces of drywall to the studs if necessary. For easier fastening, rest the bottom of the sheet on a ¼-inch- to ½-inch-thick shim or scrap to leave a gap at the floor. A trim board or baseboard will cover the gap later.

Step 2. Fasten the drywall to the framing, driving a screw (or nail) every 6 inches on the edges and every 12 inches on any inside studs.

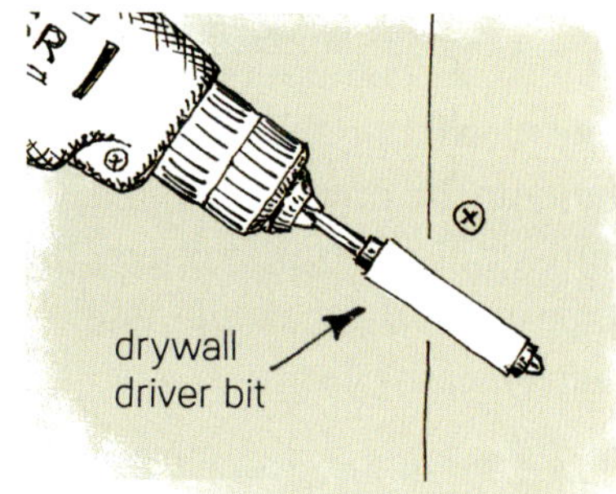

HOW TO FINISH DRYWALL

After the drywall is up, you'll want to plaster the joints and the indentations, or dimples, made by the screws or nails. You'll need some drywall knives, a roll of mesh-style joint tape, a bucket of drywall joint compound, and a mud tray.

Step 1. Check all the screws or nails to make sure the heads are slightly "dimpled" into the drywall. Then lay down self-sticking 2-inch-wide mesh tape over the joints, and press it evenly into inside corner joints.

Step 2. Open the bucket of joint compound and stir it. Use a 3-inch-wide knife to scoop out some joint compound, or mud, and plop it into the tray. While holding the tray in one hand, spread joint compound over the tape with a 6-inch knife, working it into the joint.

Step 3. Using a 9- or 10-inch knife, smooth out the compound in long, firm strokes, removing ridges, air holes, and other irregularities. Feather out the compound at its edges. Let the compound dry fully.

Step 4. Add a thin finish coat of compound over the first layer and let it dry. Repeat with a third coat. Use your 3-inch knife to cover all the nail or screw heads with compound, applying at least two layers.

Step 5. After the final layer of compound dries, sand it to remove any ridges or rough spots. When it all looks smooth enough to you, it's done.

One time-saving note: To avoid a lot of masking, prime and paint the drywall *before* you add any trim boards, paneling, or finish flooring.

Finishing Drywall

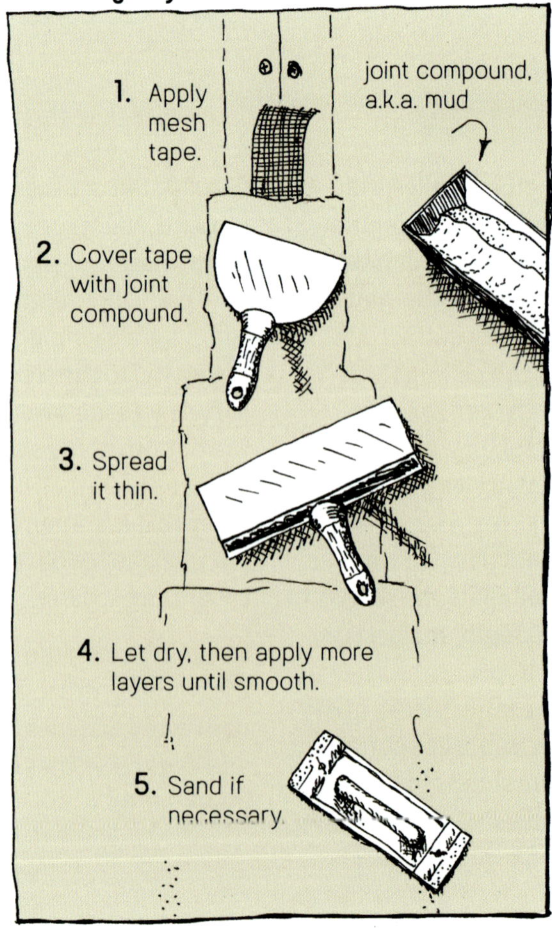

Finishing the Floor and Trim

You can sand and paint the rough flooring if you like (see Chapter 7). Or you can install finished flooring, following the manufacturer's instructions.

If you want a reasonably priced, resilient, and easy-to-install finished floor, I suggest vinyl plank flooring. You don't need any glue or special tools, and the planks can be laid directly over your OSB subfloor, no matter how rough that has become during your construction. A few area rugs will warm things up and add some splashes of color.

TRIMMING IT OUT

After your floor is finished, you can add baseboards along the floor and trim boards around your door and windows. The simplest trim method is to use 1×3 or 1×4 pine boards, which will look good with your pine paneling.

HOW TO INSTALL TRIM

Step 1. Cut the boards to size as needed. The simplest way is to cut the boards square at the ends, but a neater, more elegant look can be had by mitering them, or cutting them at a 45-degree angle. For easy square or miter cuts, use a miter box and your handsaw.

Step 2. Plane and sand the long exposed edges to make them splinter-free and give them a finished look.

Step 3. With your smaller hammer, nail in the trim with 6d or 8d finish nails. Begin around the door, since the baseboard ends will later meet the sides of the door trim. Countersink the nailheads with your nail set.

Step 4. Cover up the sunken nailheads with wood filler. (You can find filler in just about any color to match your wood.) Use a putty knife to fill in the holes, then rub off any excess with a rag.

Prepaint the Trim

If you are going to paint your trim a different color than the walls, prime and paint all the trim boards on your sawhorses before nailing them in. Then you only need to touch up the nailheads, thus avoiding a lot of messy masking tape.

Enjoy!

Your shelter is finished . . . finally! Now it's time for you to reap the benefits of all you've accomplished with your backyard build. I hope you enjoyed the actual work of creating your shelter. I'm sure there were some difficult or confusing moments at times (the trademark of any good do-it-yourself project!), but I truly hope the instructions and pictures in this manual have been helpful.

This is the step-by-step guide that I wish I'd had myself when I was attempting to build my very first shelter—a clubhouse. Many sheds, room additions, and cabins later, I took on the much bigger challenge of building my own house in Oregon. That project took many months of planning and designing that involved plumbing, electrical wiring, and a masonry chimney, as well as submitting plans for my building permit. Fortunately, by then, I had also studied some good reference books and worked with a few other carpenters and builders. The house got built!

I have found that the real learning of any hands-on craft comes from the actual practice of it. Measuring, sawing boards, pounding nails, and fitting pieces of wood together all take time to master, and confidence is gained through the process. I'm so glad you've been patient with the process to get this far. If you want to now add a partition for a tiny bathroom, closet, or private alcove, frame it as you did the outside walls and cover it with pine or drywall. You can also install or buy cabinets and shelving, add a nice rug, and maybe include a deck, patio, or hot tub outside. So many possibilities!

If you are considering tackling a larger house—complete with plumbing, electrical, and heating/air-conditioning systems—then by all means, investigate the many fine reference books and videos that detail the installation of these systems. My favorite is *How to Build a House* by Larry Haun (published by Habitat for Humanity and Taunton Press). Habitat for Humanity has overseen the construction of hundreds of thousands of modest homes across the country, all with the help of nonprofessional owner-builders and other volunteers. Check out their website, habitat.org, for more inspiration and information.

Now I feel that I've done my job! It's time for you to enjoy all that you've accomplished with your backyard build, and you certainly deserve it. Put out the patio chairs and call me when the hot tub is ready.

My Oregon Coast home, built 1981–1984

GLOSSARY

barn sash: An inexpensive wood-framed window with one, four, or six panes, used for sheds or barns. They are perfect for backyard dwellings.

beam: A horizontal piece of wood that supports some part of a house or structure.

blocks or blocking: Short lengths of 2×4 that are used to hold in windows or doors, to strengthen a frame, or to provide something to nail to.

board-and-batten: A type of siding where wide boards are applied vertically to the wall framing. The battens (bats for short) are narrow boards or wooden strips that are then nailed over the joints of the wide boards.

braces or bracing: Temporary boards that hold wall framing in place once it has been plumbed.

butt: The bottom edge of a shingle or clapboard.

car siding: Also known as rustic plank or V rustic siding. Pine boards 6 to 8 inches wide that are tongue-and-grooved and suitable for interior or exterior paneling or siding. Originally used for lining the inside walls of railway freight cars.

centerline: A mark to show where the middle of a board should be placed or nailed.

clapboards: Also known as beveled or lap siding. Long, narrow boards made of pine, redwood, or cedar that are thicker at one edge than the other. Clapboards are applied horizontally and overlapped to cover the outer walls of a wood-framed house.

coated sinker: A common steel nail that comes in sizes from about 1 inch long (3d) to 3½ inches long (16d). It is coated with a lubricant that allows the nail to be easily driven in but then immediately turns sticky to hold the nail tight in the wood.

cripple: A short stud below or above a window or door.

drywall: Also called Sheetrock, gypsum board, or wallboard. A common interior wall panel made of gypsum, a fireproof, plasterlike mineral, and covered with heavy paper. Drywall is screwed or nailed to the studs and then can be plastered, painted, wallpapered, etc.

duplex nail: A common steel nail with two heads, used for temporary bracing. When the first head is all the way in, the second head is still far enough out to allow the nail to be pulled out with a hammer.

flush: Having surfaces on the same plane; even.

footprint: The area of ground a structure covers or occupies, usually described in square feet.

frame or framing: *(noun)* The wooden skeleton of a building made up of studs, joists, rafters, etc. *(verb)* To assemble the frame of a building.

furring strips: Strips of wood, usually sold in 8-foot lengths, that are installed over framing for attaching paneling, drywall, or other finish materials. Furring strips are usually cheaper than "normal" lumber but are of similar quality.

header: A beam that supports the wall or roof above a door or window.

jamb: One part of the frame enclosing a door or a window.

joists: Boards laid parallel to each other and extending from wall to wall to support a floor or ceiling. Joists are usually laid 16 to 24 inches on center, meaning 16 or 24 inches from the center of one joist to the center of the next.

level: The condition of being perfectly horizontal, which can be tested with the tool that is also called a level.

lumber: Pieces of wood that are sawn, most commonly, from the logs of pine, fir, hemlock, cedar, or redwood trees. Lumber mills near forests saw the logs into standard-size boards that are then graded and sold through lumberyards and home centers.

nailer: An extra stud, rafter, or joist that provides a surface on which to nail sheathing, siding, paneling, or drywall.

oriented strand board (OSB) or waferboard: An inexpensive 4 × 8-foot panel made of wood chips (strands or wafers) laid down in a random (oriented) pattern, then glued and pressed together under tons of pressure. This material is used for sheathing or siding on houses all over the world.

paneling: Any thin plywood, fiberboard sheet, or boards used to finish an interior wall.

partition: An interior wall that divides a large room into smaller rooms.

paver: Any brick, stone, or concrete block used for walkways, patios, or driveways. Pavers come in a variety of colors, shapes, and sizes.

pier: A masonry (brick or concrete) column that holds up a main beam or a sill under the first floor.

pier block: A specially shaped concrete block with a wide bottom and narrow top used to hold up a foundation post or sill.

pilot hole: A hole drilled or poked into wood to accommodate or guide a screw. The pilot hole is slightly narrower than the shaft of the screw to allow the screw to "bite" into the wood.

pitch: The slope, or steepness, of a roof. A 1:12 pitch is almost flat; a 4:12 pitch is steeper, but you can still walk on it; and a 12:12 pitch is very steep.

plate: The top or bottom board that holds all the studs together in a wall frame. A basic wall frame has one bottom plate and one top plate. On exterior walls, a second top plate is added to tie the walls together.

plumb: The condition of being exactly vertical, or straight up and down, which can be tested with a level or a plumb bob.

plumb bob: A pointed conical-shaped weight held at the end of a string to test a wall or a post for plumb.

pressure-treated: Wood that has had preservative chemicals forced into the grain under high pressure. Pressure-treated wood is used mostly for exposed sills or beams, decking, fence posts, and other structural elements that may get wet or touch the ground.

rafters: Sloping boards laid parallel to each other to frame or support a roof.

rim joists: Joists that are set on the edge, or rim, of the foundation and run perpendicular to the floor joists.

sheathing: The structural exterior covering of a building, usually consisting of boards, plywood, or oriented strand board, that often adds to the building's strength and is then painted or covered with siding.

shim: A tapered, shingle-like wedge of wood used to level some part of a structure or to fill in a gap to hold something in place.

siding: Any exterior covering on a building, usually applied over sheathing. Common wood siding materials are clapboards, shingles, board-and-batten, and T1-11 plywood. Other common materials are vinyl siding, aluminum siding, fiber-cement board, brick, stone, and stucco.

sill: The beam or flat board laid on posts, piers, or the foundation wall that in turn supports the entire building.

square: A building is square when all the corners are at right angles, or 90 degrees. This includes any rectangular or square-shaped room or building. A square foundation will make the rest of the building much easier to build.

stucco: An exterior plaster made of Portland cement, sand, and water, with acrylic binder and fiberglass sometimes added. Stucco is usually applied over a plastic or metal mesh lath.

stud: An upright post in the frame of a wall. Studs support the siding on the exterior and the drywall or paneling on the interior, and they also support the roof or upper stories. Studs are usually 2×4s or 2×6s.

T1-11 plywood: A type of plywood siding that also serves as structural sheathing. The sheets are smooth or have grooves every 4, 8, or 12 inches for a design effect.

toenailing: A method of nailing thick boards together at an angle. (See the illustration on page 48.)

tongue-and-groove: An interlocking system that holds boards tightly together. Each board has a "tongue" on one edge that fits into the groove of its neighbor so that together they can take a lot of stress, such as in the floor of a basketball court.

trimmer: A block attached to the stud adjoining a window or door to strengthen the framing.

METRIC CONVERSIONS

LENGTH

TO CONVERT	TO	MULTIPLY
inches	millimeters	inches by 25.4
inches	centimeters	inches by 2.54
inches	meters	inches by 0.0254
feet	meters	feet by 0.3048
feet	kilometers	feet by 0.0003048
yards	centimeters	yards by 91.44
yards	meters	yards by 0.9144
yards	kilometers	yards by 0.0009144
miles	meters	miles by 1,609.344
miles	kilometers	miles by 1.609344

US	METRIC
⅛ inch	3.2 mm
¼ inch	6.35 mm
⅜ inch	9.5 mm
½ inch	1.27 cm
⅝ inch	1.59 cm
¾ inch	1.91 cm
⅞ inch	2.22 cm
1 inch	2.54 cm

US (INCHES)	METRIC (CENTIMETERS)
0.5	1.27
1	2.54
1.5	3.81
2	5.08
2.5	6.35
3	7.62
3.5	8.89
4	10.16
4.5	11.43
5	12.70
5.5	13.97
6	15.24
6.5	16.51
7	17.78
7.5	19.05
8	20.32
8.5	21.59
9	22.86
9.5	24.13
10	25.40
11	27.94
12	30.48
13	33.02

US (INCHES)	METRIC (CENTIMETERS)
14	35.56
15	38.10
16	40.64
17	43.18
18	45.72
19	48.26
20	50.80
21	53.34
22	55.88
23	58.42
24	60.96
25	63.50
26	66.04
27	68.58
28	71.12
29	73.66
30	76.20
31	78.74
32	81.28
33	83.82
34	86.36
35	88.90
36	91.44

Index

F

G

H

I

J

K

L

M

N

O

P

R

S

T

U

W

X